Adversarial
Risk Analysis

Adversarial Risk Analysis

David L. Banks
Duke University
Durham, North Carolina, USA

Jesus Rios
IBM Thomas J. Watson Research Center
Yorktown Heights, New York, USA

David Ríos Insua
Institute of Mathematical Sciences, ICMAT-CSIC
Madrid, Spain

CRC Press
Taylor & Francis Group
Boca Raton London New York

CRC Press is an imprint of the
Taylor & Francis Group, an **informa** business
A CHAPMAN & HALL BOOK

CRC Press/Chapman & Hall
Taylor & Francis Group
6000 Broken Sound Parkway NW, Suite 300
Boca Raton, FL 33487-2742

First issued in hardback 2019
First issued in paperback 2021

ISBN 13: 978-1-03-209849-4 (pbk)
ISBN 13: 978-1-4987-1239-2 (hbk)

Contents

Preface

A few years ago, while planning this book, we were strolling the streets of Segovia, and stopped in at the Alcázar. It is now a museum, but in previous centuries it was fortress, a state prison, a royal castle, a military academy, and the home of the Royal Artillery College. While examining the exhibits we saw a monument in honor of Captain Luis Daoíz y Torres and Captain Pedro Velarde y Santillán, two heroes of the Spanish War of Independence and alumni of the Royal Artillery College. The monument had a motto proposed by Luis de Góngora:

Science Wins.

We all agreed it was the perfect quote for the preface to our book.

The field of adversarial risk analysis is fairly young. It was prompted by the international response to the terrorist attacks of September 11, 2001, in which there was massive investment, by many countries, in measures to reduce vulnerability. In the United States, the federal government launched two invasions, established a new federal agency, and spent large sums to defend against bioterrorism, defend against nuclear terrorism, defend against chemical attack, secure the border, screen air travel passengers, improve counterintelligence, and protect federal buildings, among many other responses.

Different people had different ideas about which investments were cost-effective. But there was no principled methodology for deciding which security measures should be adopted. Some wanted to invest in almost every conceivable defense, but that would have been ruinously expensive. Some wanted to choose according to a game-theoretic perspective, while others preferred to use probabilistic risk analysis.

In this confusion, we began to think about how such decisions might be made. Ultimately, we felt that a decision maker should build a statistical model for the strategic thinking of the opponent, expressing his full uncertainty through Bayesian distributions over all unknowns. This structure enables the decision maker to select the action that maximizes his expected utility.

The chief problem, of course, is to model the reasoning of the opponent. Irresistibly, one is reminded of the dialogue from *The Princess Bride*:

Vizzini: But it's so simple. All I have to do is divine from what I know of you: are you the sort of man who would put the poison into his own goblet or his enemy's? Now, a clever man would put the poison into his own goblet, because he would know that only a great fool would reach for what he was given. I am not a great fool, so I can clearly not choose the wine in front of you. But you must have known I was not a great fool, you would have counted on it, so I can clearly not choose the wine in front of me.
Man in Black: You've made your decision then?
Vizzini: Not remotely. Because iocane comes from Australia, as everyone knows, and Australia is entirely peopled with criminals, and criminals are used to having people not trust them, as you are not trusted by me, so I can clearly not choose the wine in front of you.
Man in Black: Truly, you have a dizzying intellect.
Vizzini: Wait til I get going! Now, where was I?

It is easy to satirize this reasoning, but strategic decision making requires one to put oneself into the mental shoes of the opponent. Perhaps quixotically, we feel that proper modeling of the opponent's thinking is the key to making appropriate investments in counterterrorism.

Our work began in the fall of 2007, as part of the research program on *Extreme Events, Decision Theory and Risk Analysis* sponsored by the Statistical and Applied Mathematical Sciences Institute in the Research Triangle Park, North Carolina. In addition, David Banks thanks DARPA for grant 56711-NS-DRP, the Institute for Homeland Security Solutions for grant #3831933, and CREATE (at the University of Southern California) for grant #3833789. And David Ríos Insua thanks the Spanish Ministry of Economy and Competitiveness for grants MTM 2011-28993 and MTM2014-56949-C3-1-R, the Riesgos-CM program of the Government of Madrid, the ESF for COST programs ALGODEC IC0602 and EJNET IS1304, the SECONOMICS FP7 grant #285223, and the AXA Research Fund for the AXA-ICMAT Chair in Adversarial Risk Analysis (through ICMAT-CSIC and FGCSIC).

Among our collaborators, David Banks thanks Shouqiang Wang and Francesca Petralia. David Ríos Insua thanks A. Tsoukias, J. C. Sevillano, A. Couce, C. Gil, J. Guerrero, P. G. Esteban, J. Cano, E. L. Cano, and S. Houmb. He is grateful for the support of M. de León, director of ICMAT-CSIC. And he misses the happy hours of collaboration with S. Rios, S. Rios Insua, and J. Muruzabal.

With their help, and the slow polish of many papers and presentations, we wrote this book. We hope you enjoy it. And we hope you find it useful.

Durham, NC, USA   DAVID L. BANKS
Yorktown Heights, NY, USA   JESUS RIOS
Valdoviño, Spain   DAVID RÍOS INSUA

January 7, 2015

Chapter 1
Games and Decisions

Recent conflicts have underscored the need for decision makers to take account of the strategies of their opponents. War, terrorism, politics, corporate competition, and government regulation are all arenas in which there are multiple agents with differing goals and each agent pursues its own interest.

The traditional approach in economics and mathematics to such problems has been game theory. Game theory is a rich discipline that has made many important contributions, but it is not adequate to the strategic and behavioral complexity of many real-world applications. The difficulties with game theory are well known; these are described in Section 1.1.

Decision analysis is an alternative to game theory. It developed from Bayesian statistics, and is founded on the premise that decision makers should minimize their expected loss, or, equivalently, maximize their expected utility. This is a fundamentally different solution concept than the minimax principle that is the basis for game theory. The minimax principle asserts that decision-makers should seek to minimize their maximum possible loss.

This monograph focuses on decision analysis, and especially the new subfield called *adversarial risk analysis* (ARA). In ARA, the decision maker builds a model for the strategy of his opponent, incorporating uncertainty through probability distributions. This model induces a distribution over the actions of the opponent, and the decision maker then chooses the option that maximizes his expected utility (or, equivalently, minimizes his expected loss).

The remainder of this chapter provides context. Section 1.1 is a review of game theory; it gives definitions, outlines some of the recent generalizations that improve its applicability, and describes the advantages and disadvantages of the game-theoretic perspective. Section 1.2 provides a parallel overview of decision analysis. Section 1.3 introduces influence diagrams, which are helpful in subsequent discussion.

1.1 Game Theory: A Review

Game theory is the branch of mathematics that finds "optimal" solutions when two or more opponents with competing interests must select among possible actions. Much of the research has focused on toy problems which illustrate important distinctions among games. Some of these key distinctions are:

- Opponents may make simultaneous or sequential decisions. Games with simultaneous decisions are sometimes called *normal form* games; in these, all players announce their decision at the same time. In contrast, players in sequential games make their decisions over time, perhaps in response to previous decisions by other players. Most commonly, two players will alternate in declaring their decisions, and these are called *extensive form* games.
- Games may be discrete or continuous. In discrete games, the set of choices is countable (and usually finite). In continuous games, the choice sets are uncountable. For example, an auction is a discrete game if the players must bid in pennies, but a continuous game if the bids may be arbitrary fractions of a penny.
- Games may be deterministic or stochastic. If all players know each other's possible actions and the consequences corresponding to every collection of choices, then it is a deterministic game. A stochastic game occurs when this strong condition fails. In particular, in many cases, the payoff resulting from a specific combination of choices by the opponents is random. (The deterministic/stochastic distinction is related to, but not the same as, the distinctions between games with perfect/imperfect information and between games with complete/incomplete information; cf. Myerson 1991, Chaps. 2.1 and 7.3.)
- Games may be zero-sum or non-zero-sum. In a zero-sum game, what one player gains another player must lose (e.g., as in gambling). In a non-zero-sum game, the total gains or losses may be more or less than zero. Non-zero-sum games allow the possibility of win-win solutions among some or all the players (e.g., competing companies might collaborate to develop a product that creates new revenue for both).
- Games may or may not allow communication between the players (or signaling). When communication is possible, then there is the potential for threats, bluffs, use of disinformation, and cooperation.
- Games may have two or more players. When there are more than two players, then there is the possibility that some of them will form coalitions, which increases strategic complexity.

In practice, these distinctions may not be so sharp. For example, the number of players can change over the course of a game, or games may evolve from a zero-sum game to a non-zero-sum game, or move from competition to cooperation.

A famous toy game is the Blotto game (cf. Bellman, 1969). It is an example of a two-person simultaneous finite deterministic zero-sum game.

Example 1.1: Colonel Blotto has six battalions that he must allocate across three battlefields. At least one battalion must be assigned to each location. His opponent, Colonel Klink, controls six battalions and must also place at least one in each location. Neither knows in advance how the other will assign his forces, but both know that, for each battlefield, the side with the larger number of battalions will win (and if both assign the same number to the same location, then there will be a draw). The winner of the Blotto game is the side that wins the majority of the battles.

The Blotto game is interesting because it may be viewed as a toy model for counterterrorism investment. A government must decide how much to invest in border security, cargo inspections, airline screening, and other defense initiatives, while the terrorists must decide how to allocate resources among the possible attacks.

Before solving a Blotto game, first consider its characteristics. There are clearly two players, and moves are simultaneous—the allocation of battalions is not known until the troops engage. The choice sets are finite: there is only a fixed number of ways that Colonel Blotto can allocate his troops, and similarly for Colonel Klink. The game is deterministic, since both opponents know how many battalions the other controls, and chance plays no role in the outcome at a battlefield (but later that assumption is relaxed). Finally, the game is zero-sum because a win at a battlefield for Colonel Blotto is a loss for Colonel Klink, and vice versa.

Both Colonel Blotto and Colonel Klink have the same choice set, which is any triple of whole numbers that sum to 6:

$$(1, 1, 4) \quad (1, 4, 1) \quad (4, 1, 1) \quad (1, 2, 3) \quad (1, 3, 2)$$
$$(2, 1, 3) \quad (2, 3, 1) \quad (3, 1, 2) \quad (3, 2, 1) \quad (2, 2, 2)$$

Note that these allocations are just permutations of $(1, 1, 4)$, $(1, 2, 3)$, and $(2, 2, 2)$.

Traditionally, the Blotto game ignores permutations of an allocation. On average, this is appropriate; and it makes this a deterministic game. But it is also possible to view this as a stochastic game. For example, suppose Colonel Blotto chooses the allocation $(1, 2, 3)$ and Colonel Klink chooses randomly among any of the permutations of $(1, 2, 3)$. Four of the possible pairs of choice lead to a tie. But if Colonel Klink chooses $(3, 1, 2)$, he wins the first battlefield and loses the other two; if he chooses $(2, 3, 1)$, he wins the first two battlefields but loses the last. If the permutation is chosen at random, the expected outcome is a draw. For Blotto games in general, with arbitrary and possibly unequal numbers of battalions apiece, one can show that randomly permuted allocations have expected value equal to the outcome of an increasing sequence allocation. For this reason, game theorists often ignore order in Blotto games.

To analyze the Blotto game, consider the payoff matrix below. The columns represent Colonel Blotto's choice set, and the rows represent Colonel Klink's. Within each cell, the first entry is the payoff to Colonel Blotto, and the second entry is the payoff to Colonel Klink. Quite unrealistically, the conventional payoff is 1 for winning the majority of battlefields, −1 for losing the majority of battlefields, and 0 for draws. (Since this is a zero-sum game, the first entry is the negative of the second.)

	Blotto		
	(1, 1, 4)	(1, 2, 3)	(2, 2, 2)
Klink (1, 1, 4)	0, 0	0, 0	1, −1
(1, 2, 3)	0, 0	0, 0	0, 0
(2, 2, 2)	−1, 1	0, 0	0, 0

The payoff matrix shows that (2, 2, 2) beats (1, 1, 4), and every other pair of choices yields a draw.

Most analysts would advise Colonel Blotto to choose (2,2,2). No other choice can win, and if there is a chance that Colonel Klink foolishly chooses (1,1,4), then Colonel Blotto will prevail. But there is an argument that Colonel Blotto should choose (1,2,3): he cannot lose if Colonel Klink plays (2,2,2) or (1,1,4), and if Colonel Klink also plays (1,2,3), then a random assignment of his troop strength to specific battlefields implies that Colonel Blotto has 1/6 chance of winning, 1/6 chance of losing, and 2/3 chance of a draw. In some situations, that might be preferable to a guaranteed stalemate.

Classical game theory sees both (2,2,2) and (1,2,3) as solutions. Formally, a pair of choices is a *Nash equilibrium* if neither player can gain by unilaterally changing his choice. This means that Colonel Blotto is making the best decision possible, taking account of Colonel Klink's decision, and symmetrically, Colonel Klink is making the best decision possible, taking account of Colonel Blotto's. For the Blotto game, all four possible pairs of choices taken from $\{(2,2,2),(1,2,3)\}$ are Nash equilibria.

For two-person zero-sum games, von Neumann and Morgenstern (1944) proved that a Nash equilibrium solution always exists.

John Nash (1951) extended this to a much larger class of games: at least one Nash equilibrium exists for any game with finite choice sets (thus allowing more than two players and non-zero-sum situations). Both proofs require the concept of *randomized strategies*.

In the Blotto example, the pair (2,2,2) and (2,3,1) jointly constitute a pure strategy Nash equilibrium, in the sense that Colonel Blotto can always play (2,2,2) and Colonel Klink can always play (2,3,1) and the conditions for a Nash equilibrium are satisfied. In contrast, with a randomized strategy, a Nash equilibrium is achieved by having Colonel Blotto play the ith allocation with some probability p_i, and Colonel Klink play the ith allocation with some probability q_i, where both sets of probabilities sum to 1. For many games, the only Nash equilibrium solutions are randomized strategies.

The game gets more complex as the number of battalions increases. When there are more than 12 battalions apiece, no pure strategy is a Nash equilibrium. For example, with 13 battalions, Colonel Blotto should choose allocation (3,5,5) with probability 1/3, allocation (3,3,7) with probability 1/3, and allocation (1,5,7) with probability 1/3, and Colonel Klink should do likewise (cf. Roberson, 2006).

The Blotto game is deliberately simplistic. One such simplification concerns the payoff: a 1 for a win, a -1 for a loss, and a 0 for a draw. In more realistic scenarios, the value of a win could be large (if it resolved the war) or small (if it were a minor skirmish). In game theory and decision analysis, one handles this valuation problem through the *utility* of an outcome. The utility combines all the costs and benefits into two numbers that summarize the net payoff to Colonel Blotto and the net payoff to Colonel Klink. This payoff could be positive or negative, and it would include such things as the cost in human lives, the military value of a victory or loss, the financial resources needed to manage the operation, and so forth. If the colonels were personally ambitious, it could even include the impact of the win or loss on their chances for promotion.

A second simplification is the assumption that the outcome is deterministic, depending only upon the number of battalions that each opponent allocates. By chance, an inferior force might defeat superior numbers, or force a draw. Also, the cost of a defeat may be small, if an orderly retreat is achieved, or large, if there were a massacre. Thus, it would be more realistic to describe the utility that is realized from a particular pair of allocations as a random variable, rather than some known quantity.

Realistic uncertainty causes other complications. For example, it is unlikely that Colonel Blotto knows exactly the utility that Colonel Klink assigns to a win, loss, or draw. And Colonel Blotto may have received intelligence regarding the allocations Colonel Klink will make—he is not certain of the accuracy of the intelligence, but should it be ignored? Finally, Colonel Blotto may not know whether Colonel Klink is selecting his allocation based on the mathematical solution to a game theory problem, or whether he is following orders from his commanding officer, or whether he is simply playing a hunch, or whether he is using some other system. In real life, all of these uncertainties are relevant to the problem. Typically, analysts attempt to express such uncertainty through probability distributions.

In game theory, the random payoffs and unknown utilities are treated together, using the *theory of types* (cf. Harsanyi, 1967a,b, 1968). In this framework, uncertainty about the payoff matrix is handled by imagining that each opponent is actually drawn at random from a population of possible opponents, and that different members of that population receive different payoffs from the same combinations of strategies. Similarly, and rather elegantly, this also accounts for different utility functions; different draws provide opponents with different utilities for the same event. The random draw determines the type of the opponent: with a certain probability, the opponent may be an expert or an amateur, and may value a success highly or modestly. Such games are called *Bayesian games*.

To solve a Bayesian game, one assumes that each opponent seeks to maximize his expected utility, and then finds the *Bayes Nash equilibria*. Any Bayes Nash equilibrium is a set of actions and a set of distributions (specified for each player) regarding the types of each of the other players, such that these sets maximize the expected utility for each player, given all their distributions on the types of their opponents and on the actions chosen by all of their opponents. This quickly becomes complicated, and may require numerical optimization to discover the solutions.

To illustrate the role of the theory of types, consider the following scenario.

Example 1.2: Apollo is a terrorist, and will choose to attack Daphne's country with either smallpox or anthrax. Daphne must decide whether to defend by stockpiling smallpox vaccine or Cipro (budget cuts preclude stockpiling both). Apollo does not know the net cost to Daphne for producing and storing the vaccine or the drug, but he has a probability distribution over that cost. Nor does Daphne know the net benefit to Apollo from either a smallpox attack or an anthrax attack, but she has probability distributions on both. We must make a strong assumption of *common knowledge*: somehow Daphne knows Apollo's distribution for her cost, and he knows her distributions for the benefit he receives from a bioterrorist attack, and both know that the other knows those distributions.

The theory of types handles the uncertainty in the payoff by allowing each player to be a random draw from a population of possible players. For example, Daphne might believe that there are two types of Apollos: those skillful enough to have a 50% chance of mounting a successful smallpox attack, and those with only a 10% chance of a successful smallpox attack. Additionally, Daphne has a subjective distribution over the type of her opponent; perhaps she thinks there is probability 0.2 that Apollo is skillful, so her probability that Apollo is unskilled is 0.8. Then the payoff is random, and depends upon, among other things, the type of Apollo whom Daphne must confront.

Similarly, different types of Apollo may value a smallpox attack versus an anthrax attack in different ways. One type might think a successful smallpox attack has more utility than a successful anthrax attack, while another type might have the opposite preference. This allows the theory of types to express uncertainty regarding the utility functions of the opponents. Thus the same technique can handle two problematic aspects of uncertainty in games.

In general, suppose Daphne can select a defense from the set $\mathcal{D} = \{d_1, \ldots, d_m\}$ and Apollo can select an attack from the set $\mathcal{A} = \{a_1, \ldots, a_n\}$. Daphne uses the decision rule $\sigma_D(x)$, where x is Daphne's type and $\sigma_D(x)$ is a (possibly randomized) choice of defense from $\mathcal{D}$. Apollo uses the decision rule $\sigma_A(y)$, where y is Apollo's type and $\sigma_A(y)$ is a (possibly randomized) choice of attack from $\mathcal{A}$. The joint distribution of types is $F(x,y)$, which allows for correlation among the types of opponents. The common knowledge assumption implies that $F(x,y)$ is known to both Daphne and Apollo.

The payoff to Daphne is determined by a utility function $u_D(\sigma_D, \sigma_A, x)$ that depends on her rule, Apollo's rule, and her true type x. Symmetrically, the payoff to Apollo is $u_A(\sigma_D, \sigma_A, y)$. A Bayes Nash equilibrium is a pair of rules (σ_D, σ_A) such that

$$E_F[u_A(\sigma_D(X), \sigma_A(Y), X)] \geq E_F[u_A(\sigma_D'(X), \sigma_A(Y), X)]$$
$$E_F[u_D(\sigma_D(X), \sigma_A(Y), Y)] \geq E_F[u_D(\sigma_D(X), \sigma_A'(Y), Y)]$$

where σ'_A is any decision rule that selects from $\mathcal{A}$ and σ'_D is any decision rule that selects from $\mathcal{D}$. At least one such solution exists, and often there are many. It is an equilibrium solution since Apollo cannot improve upon his expected payoff by unilaterally changing his rule, nor can Daphne.

It may seem worrisome that the decision makers are not conditioning on their true types. Daphne knows that her type is $X = x$, and Apollo knows that his true type is $Y = y$. But it turns out that the Bayes Nash equilibrium above also solves the analogous conditional expectations:

$$E_{F(y|X=x)}[u_D(\sigma_D(x), \sigma_A(Y), x) \,|\, x] \geq E_{F(y|X=x)}[u_D(\sigma'_D(x), \sigma_A(Y), x) \,|\, x]$$
$$E_{F(x|Y=y)}[u_A(\sigma_D(X), \sigma_A(y), y) \,|\, y] \geq E_{F(x|Y=y)}[u_A(\sigma_D(X), \sigma'_A(y), y) \,|\, y]$$

where $F(y|X=x)$ is the conditional distribution of Y given $X = x$, and similarly for $F(x|Y=y)$. Both conditional distributions are straightforwardly obtained from the joint distribution $F(x, y)$.

When there are multiple Bayes Nash equilibria, as often happens, the analyst must decide which one to use. Some solutions are unreasonable, according to various criteria that correspond to common sense or mathematical principles. Consequently, this area of game theory has been richly elaborated, as researchers attempt to find solution concepts that exclude unwanted equilibria. See Myerson (1991, Section 2.8) for more details.

In sequential games, the problematic assumption of common knowledge can often be removed. One example is the well-known Defend-Attack game. Brown, Carlyle and Wood (2008) provide a simple introduction to it; Alderson et al. (2011) extend the framework to more complex problems; and Bier and Azaiez (2009) give a book-length treatment of this game in the context of national security.

In a sequential Defend-Attack game, the Defender (Daphne) makes the first move. She must choose an action from the set $\mathcal{D} = \{d_1, \ldots, d_m\}$. The Attacker (Apollo) observes Daphne's choice and then selects an action from the set $\mathcal{A} = \{a_1, \ldots, a_n\}$. For each pair of choices (d_i, a_j), the outcome is a random variable S, which might be an indicator of success or failure, or a continuous loss or gain. Daphne and Apollo may have different probability distributions for the chance of success given a pair of choices (d_i, a_j). They may also have different utility functions.

Example 1.3: Daphne must decide whether or not to install millimeter wave body scanners at airport security checkpoints, replacing X-ray scanners. Apollo then decides whether or not to attempt to smuggle a bomb onto an airplane. Daphne makes the first move, so Apollo can see whether or not body scanners are in use when he arrives at the airport. Since Apollo gets to observe Daphne's action before selecting his own, he does not need to know her probabilities or her utilities. But Daphne must have a distribution on Apollo's type, which specifies his utilities and probabilities.

Fig. 1.1 The Defend-Attack
game tree ($m = n = 2$).

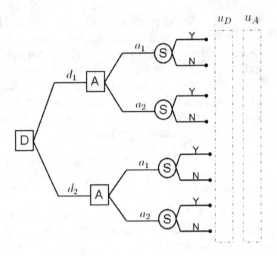

The sequential Defend-Attack game can be described through a game tree. Figure 1.1 shows the game tree for this example, in which both Daphne and Apollo have exactly two possible actions (i.e., $m = n = 2$). The squares indicate points at which decisions are made, with Apollo and Daphne distinguished by their initials. The analysis moves from left to right; Daphne chooses first, then Apollo, then they observe the random outcome, denoted by a circle. Their utilities are indicated by the dashed rectangles.

To solve the Defend-Attack game, one needs the expected utilities of players at node Ⓢ of the tree in Fig. 1.1. The expected utility that Apollo obtains when the decisions are $(d, a) \in \mathcal{D} \times \mathcal{A}$ is

$$\psi_A(d,a) = p_A(S=0 \mid d,a)\, u_A(d,a,S=0) + p_A(S=1 \mid d,a)\, u_A(d,a,S=1), \quad (1.1)$$

and similarly for Daphne's expected utility $\psi_D(d,a)$. Thus, Apollo's best attack against the defense d is

$$a^*(d) = \operatorname{argmax}_{a \in \mathcal{A}} \psi_A(d,a), \quad \forall d \in \mathcal{D}.$$

Under the assumption that Daphne knows how Apollo will solve his problem, her best choice is

$$d^* = \operatorname{argmax}_{d \in \mathcal{D}} \psi_D(d, a^*(d)).$$

The solution $(d^*, a^*(d^*))$ is a Nash equilibrium (Aliprantis and Chakrabarti, 2010, Chap. 4).

In practice, Daphne does not know how Apollo will solve his problem. She might believe that he will solve an equation of the form (1.1), but she does not know the specific u_A and p_A, the utilities and probabilities, that he will use. In such circumstances, the Bayes Nash equilibrium perspective maintains that Daphne should use her distribution over Apollo's type to impose a distribution over his u_A and p_A.

Communication (or signaling) is a critical issue in game theory. Early work on noncooperative games assumed that opponents do not communicate, exacerbating the implausibility of the common knowledge assumption. More recent work has shown that communication can expand the set of actions and reduce the chance of finding socially inferior equilibria (Spence, 1974; Cho and Kreps, 1987). Communication also enables collusion among corporate competitors for mutual benefit, even in a market designed to promote competition. And there are games in which a player is better off without communication; e.g., if the Attacker can communicate a threat, it may force a concession from the Defender without requiring the Attacker to expend any resources (cf. Raiffa, Richardson and Metcalfe, 2002).

Another key concept in modern research is that of *bounded rationality*. Simon (1955) argued that humans are usually unable to calculate the optimal solutions sought by classical game theory or decision analysis. We are not smart enough nor fast enough nor sufficiently self-consistent to solve game theory problems that arise in real interactions. In Simon's terminology, people are *satisficers*, settling for solutions that are sufficiently satisfactory and easy to find. Kahneman (2003) reviews a large body of research on bounded rationality, documenting various cognitive deficiencies and biases. Rubinstein (1998) describes various models of decision making that incorporate specific kinds of bounded rationality, such as constraints on memory, bounds and costs on calculation, and limited foresight.

Limited foresight is especially important. In game theory, it led to level-k models, which posit:

- a level-0 opponent does not use any strategy in choosing an action;
- for $k > 0$, a level-k opponent chooses the best response to opponents of level $k - 1$.

A level-k model indicates how deeply a player thinks in forming his strategy. There is empirical evidence that level-k models (where k is 1 or 2 for most situations) provide more accurate descriptions of real-world decision making than do Nash equilibrium solutions (cf. Stahl and Wilson, 1994, 1995; Nagel, 1995).

From a policy standpoint, game theory solutions can imply societal dilemmas. Heal and Kunreuther (2007) study security measures in networks of trust; specifically, they consider baggage screening by a consortium of airlines. In their example, each airline is able to screen luggage at the time it is checked; but once the bag is in transit, it will pass unexamined to other airlines when the passenger transfers. Thus the airlines trust each other to be responsible for security at the initial check-in. The overall security in the network increases if each member of the consortium improves its own screening technology. However, each individual airline is better off if it secretly defects, while the other airlines invest. The defector spends nothing and enjoys the benefit of better protection as a free ride on the investments of other network members. But if everyone defects, the result is worse for each player than if all were to cooperate. For this reason, third-party regulators impose mechanisms that ensure security investment by all. Creating such mechanisms is difficult; complex games have multiple Nash equilibria, and so regulators should expect unintended outcomes.

However, the main concern about the relevance of game theory is the large body of empirical evidence which shows that humans simply do not behave as its theory predicts. We are usually irrational, and inconsistently competitive. Also, in many situations there is insufficient common knowledge to support the standard solutions. Camerer (2003) and Gintis (2009) summarize much of this evidence, and criticize game-theoretic solutions as poor descriptions of real-world decision making.

1.2 Decision Analysis: An Introduction

Decision analysis offers a Bayesian alternative to game theory. In the context of strategic opponents, it was simultaneously proposed by Raiffa (1982) and Kadane and Larkey (1982). Its solution concept holds that players should minimize their expected loss, instead of minimizing their maximum loss. (Equivalently, players should seek to maximize their expected utility, instead of trying to maximize their minimum utility.)

Decision analysis is Bayesian in that it requires the agents to have probability distributions over the actions of their opponents. In most settings, there is no possibility of repeated independent trials, so the frequentist definition of probability is untenable. Therefore, decision analysts typically use personal probabilities, in the sense developed by Savage (1954), and these personal probabilities describe their subjective beliefs about the likely actions of the other agents.

Decision analysis is controversial. Roger Myerson, a 2007 Nobel laureate in economics, gives a concise criticism:

> A fundamental difficulty may make the decision-analytic approach impossible to implement, however. To assess his subjective probability distribution over the other players' strategies, player *i* may feel that he should try to imagine himself in their situations. When he does so, he may realize that the other players cannot determine their optimal strategies until they have assessed their subjective probability distributions over *i*'s possible strategies. Thus, player *i* may realize that he cannot predict his opponents' behavior until he understands what an intelligent person would rationally expect him to do, which is, of course, the problem that he started with. This difficulty would force *i* to abandon the decision analytic approach and instead undertake a game-theoretic approach, in which he tries to solve all players' decision problems simultaneously. [Myerson 1991, pp. 114–115.]

Myerson's quick dismissal overlooks the practical advantages of decision analysis in modeling human behavior. Humans rarely frame problems in the way he describes.

John Harsanyi, one of the 1994 Nobel laureates in economics, also objected to decision analysis. In his discussion of Kadane and Larkey (1982), he argued that decision analysis was contrary to the spirit of game theory, arguing that probability assessment of the adversaries' actions should be based upon an analysis of their rational behavior (Harsanyi, 1982). Harsanyi implies that these probability assessments should not be based upon their actual behavior in similar games, nor upon judgments about an adversary's knowledge, motives, or strategic acuity. As previously noted, Harsanyi's alternative solution concept, the Bayes Nash equilibrium, requires impractical assumptions about common knowledge.

In decision analysis, one considers the problem from the standpoint of a single agent, using only his beliefs and knowledge, rather than trying to solve all of the agents' problems simultaneously. The selected agent is assumed to have

- a subjective probability about the actions of each opponent,
- subjective conditional probabilities about the outcome for every set of possible choices, and
- perfect knowledge of his own utility function.

Specifically, in a two-person simultaneous game, Daphne might believe that Apollo has probability $\pi_D(a)$ of choosing action $a \in \mathcal{A}$. She also assigns subjective probability $p_D(s|a,d)$ for each possible outcome $s \in \mathcal{S}$ given every choice $(a,d) \in \mathcal{A} \times \mathcal{D}$. And she knows her own utility $u_D(d,a,s)$ for each combination of outcome and pair of choices. Within that framework, Daphne maximizes her expected utility by choosing the action d^* such that

$$d^* = \operatorname{argmax}_{d \in \mathcal{D}} \mathbb{E}_{\pi_D, p_D}[u_D(d,A,S)] \tag{1.2}$$

$$= \operatorname{argmax}_{d \in \mathcal{D}} \int_{s \in \mathcal{S}} \int_{a \in \mathcal{A}} u_D(d,a,s) p_D(s|d,a) \pi_D(a) \, da \, ds,$$

where A is the random action chosen by Apollo and S is the random outcome that results from choosing A and d. The integral is interpreted as a sum when the relevant space is discrete.

Example 1.1, continued: Suppose that Colonel Blotto believes that Colonel Klink has probability $\pi_B(x)$ of choosing the action $x = (x_1, x_2, x_3)$, which indicates that Colonel Klink assigns x_i battalions to battlefield i, where the x_i are positive integers that sum to 6. Also, Colonel Blotto believes that if he assigns y_i of his battalions to location i, and if Colonel Klink assigns x_i battalions to location i, then the conditional probability that Colonel Blotto wins in battlefield i is $p_B(S_i = 1 | x_i, y_i)$. Finally, Colonel Blotto has a utility function $u_B(\mathbf{x}, \mathbf{y}, \mathbf{s})$, where $\mathbf{s} = (s_1, s_2, s_3)$ is the set of military outcomes, success or defeat, and $\mathbf{y}$ is the theater allocation chosen by Colonel Blotto.

In this situation, it could be that Colonel Blotto has no insight into Colonel Klink's strategy, and thus puts equal probability on all possible assignments. Or, it could be that Colonel Blotto believes that Colonel Klink has studied game theory, in which case he might put probability 1 on assignment $(2,2,2)$ (or perhaps he puts probability 1/6 on each of the six permutations of $(1,2,3)$, as previously discussed). But it is more interesting to suppose that Colonel Blotto has received a tip from an informant which leads him to believe that Colonel Klink has probability 1/2 of making the assignment $(1,2,3)$, and equal probability, 1/18, of making any other of the nine remaining assignments. These probabilities play the role of the $\pi_D(a)$ needed for the decision analysis in (1.2).

At a particular battlefield, Colonel Blotto's troops win if $y_i > x_i$ and lose if $y_i < x_i$, so a numerical advantage is decisive. But in the event that $y_i = x_i$, Colonel Blotto may believe that his forces have an edge (perhaps this belief is based upon previous combat experience, or information about morale and resources). So when both opponents assign an equal number of battalions to the same location, Colonel Blotto thinks that his probability of winning is 5/10, his probability of losing is 1/10, and his probability of a draw is 4/10. These values determine the $p_B(S_i | x_i, y_i)$ needed for the decision analysis.

Finally, we suppose that Colonel Blotto uses the traditional utility function. He gains 1 unit if he wins a majority of the battlefields, he loses 1 unit if Colonel Klink wins the majority of the battlefields, and the gain is 0 if both win the same number of battlefields. This determines the utility function $u_B(x, y, s)$, which plays the role of the $u_D(a, d, s)$ used in (1.2).

Tables 1.1 and 1.2 organize the decision analysis calculation. The rows in Table 1.1 correspond to the allocations chosen by Colonel Klink, and the columns are Colonel Blotto's possible allocations. Table 1.1 also shows the $\pi_B(x)$ as the second column from the right. Colonel Blotto wants to select the column which has the largest expected utility.

For a given pair of allocations, the corresponding cell in the table indicates, for each battlefield, whether Colonel Blotto has more battalions (W, for win), fewer battalions (L, for loss) or the same number of battalions (T, for tie). So if Colonel Klink selects (123) and Colonel Blotto selects (321), then the cell contains (WTL). (Without confusion, the commas in the ordered triplets have been omitted to save space.)

If the outcome is (WWL), then the utility for Colonel Blotto is 1. If the outcome is (LLW), then the utility is -1. But if there are ties, then it is necessary to calculate the expected utility in that cell using the personal probabilities that Colonel Blotto has for victory, loss and draw in a tied battlefield—these play the role of the $p_D(s | d, a)$ probabilities described previously. As Table 1.1 shows, there are only two cases to consider: those with exactly one L, one W, and one T, and the case where all battlefields receive the same allocation, (TTT).

Case (TWL): Colonel Blotto wins if his troops prevail when an equal number of battalions is assigned, which has probability 0.5. He loses with probability 0.1, and draws with probability 0.4. Thus the expected utility in this cell is $(0.5)(1) + (0.1)(-1) + (0.4)(0) = 0.4$. The same analysis applies to any permutation of T, W, and L.

Case (TTT): Colonel Blotto succeeds if his troops win two or more of the battles: this happens with probability $3(0.5)^2(0.5) + (0.5)^3 = 0.5$. He fails if his troops lose two or more of the battles: this happens with probability $3(0.1)^2(0.9) + (0.1)^3 = 0.028$. All other outcomes yield a draw. Thus Colonel Blotto's expected utility in that cell is $(0.5)(1) + (0.028)(-1) + (0.472)(0) = 0.472$, which rounds to 0.47.

Table 1.1 This table shows, for each possible pair of allocations, whether the result will lead to numerical superiority, numerical inferiority, or a tie in each battlefield for Colonel Blotto. The probabilities Colonel Blotto has for Colonel Klink's allocations are shown in the second column.

Klink	$\pi_B(x)$	Blotto's Choices									
		(123)	(114)	(141)	(411)	(132)	(213)	(231)	(312)	(321)	(222)
(123)	1/2	(TTT)	(TLW)	(TWL)	(WLL)	(TWL)	(WLT)	(WWL)	(WLL)	(WTL)	(WTL)
(114)	1/18	(TWL)	(TTT)	(TWL)	(WTL)	(TWL)	(WTL)	(WWL)	(WTL)	(WWL)	(WWL)
(141)	1/18	(TLW)	(TLW)	(TTT)	(WLT)	(TLW)	(WLW)	(WLT)	(WLW)	(WLT)	(WLW)
(411)	1/18	(LWW)	(LTW)	(LWT)	(TTT)	(LWW)	(LTW)	(LWT)	(LTW)	(LWT)	(LWW)
(132)	1/18	(TLW)	(TLW)	(TWL)	(WLL)	(TTT)	(WLW)	(WTL)	(WLT)	(WLL)	(WLT)
(213)	1/18	(LWT)	(LTW)	(LWL)	(WTL)	(LWL)	(TTT)	(TLW)	(WTL)	(WWL)	(TWL)
(231)	1/18	(LLW)	(LLW)	(LWT)	(WLT)	(LTW)	(TWL)	(TTT)	(WLW)	(WLT)	(TLW)
(312)	1/18	(LWW)	(LTW)	(LWL)	(WTL)	(LWT)	(LTW)	(LWL)	(TTT)	(TWL)	(LWT)
(321)	1/18	(LTW)	(LLW)	(LWT)	(WLT)	(LWW)	(LLW)	(LWT)	(TLW)	(TTT)	(LTW)
(222)	1/18	(LTW)	(LLW)	(LWL)	(WLL)	(LWT)	(TLW)	(TWL)	(WLT)	(WTL)	(TTT)

Table 1.2 Each cell in the body of the table shows Colonel Blotto's expected utility for that combination of allocations chosen by himself and Colonel Klink. The last row of the table shows the overall expected utility among his ten options, based on the probabilities held by Colonel Blotto.

Klink	$\pi_B(x)$	Blotto's Choices									
		(123)	(114)	(141)	(411)	(132)	(213)	(231)	(312)	(321)	(222)
(123)	1/2	0.47	0.40	0.40	−1.00	0.40	0.40	1.00	−1.00	0.40	0.40
(114)	1/18	0.40	0.47	0.40	0.40	0.40	0.40	1.00	0.40	1.00	1.00
(141)	1/18	0.40	0.40	0.47	0.40	0.40	1.00	0.40	1.00	0.40	1.00
(411)	1/18	1.00	0.40	0.40	0.47	1.00	0.40	0.40	0.40	0.40	1.00
(132)	1/18	0.40	0.40	0.40	−1.00	0.47	1.00	0.40	0.40	−1.00	0.40
(213)	1/18	0.40	0.40	−1.00	0.40	−1.00	0.47	0.40	0.40	1.00	0.40
(231)	1/18	−1.00	−1.00	0.40	0.40	0.40	0.40	0.47	1.00	0.40	0.40
(312)	1/18	1.00	0.40	−1.00	0.40	0.40	0.40	−1.00	0.47	0.40	0.40
(321)	1/18	0.40	−1.00	0.40	0.40	1.00	−1.00	0.40	0.40	0.47	0.40
(222)	1/18	0.40	−1.00	−1.00	−1.00	0.40	0.40	0.40	0.40	0.40	0.47
Expected Utility		0.47	0.17	0.17	−0.45	0.39	0.39	0.66	−0.23	0.39	0.50

Table 1.2 inserts the expected cell values into Table 1.1, and adds a final row that shows Colonel Blotto's expected utility for each of his possible choices (i.e., the dot product of the second column with the column of utilities for each choice). Clearly, Colonel Blotto will maximize his expected utility (0.66) by selecting the allocation (2,3,1).

The main difficulty in implementing a decision analysis is to determine the subjective beliefs held by the agent regarding his opponent's actions. For this example, Colonel Blotto's probabilities were based upon a tip from an informant (perhaps after making subjective adjustments to account for the reliability of the informant). When such intelligence is available, it is a valuable input to the decision analysis.

However, credible intelligence is often either not available or not sufficient to determine the $\pi_D(a)$. In that case, one may want to appeal to some kind of game-theoretic reasoning. But traditional decision analysis has not been explicit on how to use a strategic perspective to construct one's personal probabilities. This issue is touched upon by Kadane (2009), in the context of Bayesian inference in early-modern detective stories:

> The main idea is that if I am playing a game against you, my main source of uncertainty is what you will do. As a Bayesian I have probabilities on what you will do, and can use them to calculate my maximum expected utility choice, which is what I should choose. [Kadane 2009, p. 243.]

Kadane's advice is correct—Bayesians always express their uncertainty through probabilities. But he gives no prescription for how these uncertainties arise from a combination of strategic analysis of the game and whatever knowledge he may have of his opponent.

To fill that gap, adversarial risk analysis (ARA) asserts that the analyst should build a model for the decision-making process of his opponent. Within that model, the analyst then solves the problem from the perspective of the opponent, while placing subjective probability distributions on all unknown quantities. ARA leads to a distribution over the actions of the opponent. That distribution then plays the role of $\pi_D(a)$ in the conventional decision analysis, enabling the analyst to maximize his expected utilities. To make this concrete, consider Example 1.2 again.

Example 1.2, continued: Apollo is deciding whether to attack Daphne's country with smallpox or anthrax. And Daphne must decide whether to stockpile smallpox vaccine or Cipro. Daphne does not know how Apollo makes his decisions, but she has a model for his process—perhaps she believes Apollo tosses a coin, where the probability of selecting an attack is proportional to the expected utility of that attack. In that case, if she knew his utilities and his probabilities, she could optimize her own selection. But since Daphne does not know his utilities and probabilities, she places her own subjective distributions over those, solves the system, and obtains her subjective distribution for the action he will take. With that distribution, she can now perform the decision analysis that determines her own choice.

Daphne's approach is called a *mirroring argument* (cf. Ríos Insua, Rios and Banks, 2009). The decision maker models the reasoning of her opponent, using subjective distributions as necessary. If the opponent is subtle, the details become complicated, as shown in Chapter 2. For now, we focus on a simple case.

First, Daphne must model how Apollo will make his choice. She might believe that Apollo is risk averse, and wants to minimize his worst-case outcome. Or she might believe that he will maximize his expected utility. She might even believe that Apollo practices ARA too, and is mirroring her analysis in order to select his best decision. In principle, Daphne could assign a subjective probability to each of these options, thus expressing her uncertainty about which kind of decision-maker Apollo is.

For specificity, suppose Daphne thinks Apollo will choose the smallpox attack with probability proportional to the expected utility he receives, and similarly for anthrax. Apollo knows that his utility from a successful smallpox attack is u_1, his utility from a failed smallpox attack is u_1^*, and he believes that his probability of success is p_1. Also, the utility from a successful anthrax attack is u_2, the utility from a failed anthrax attack is u_2^*, and his probability of success is p_2. Thus his expected utility from a smallpox attack is $x_1 = p_1 u_1 + (1 - p_1) u_1^*$ and his expected utility from an anthrax attack is $x_2 = p_2 u_2 + (1 - p_2) u_2^*$. (We assume both expected utilities are positive; otherwise, the problem is trivial). In this framework, Daphne believes Apollo will choose the smallpox attack with probability $x_1/(x_1 + x_2)$. Note that in this unrealistically simple model of an analysis, Apollo's choice ignores Daphne's action.

Daphne's model for Apollo may seem like a strange decision rule. Harsanyi would advise her to find a Bayes Nash equilibrium; Kadane would tell her to maximize her expected utility. Nonetheless, this rule is popular in the decision analysis literature within the defense community. Paté-Cornell and Guikema (2002) proposed the rule and won the Best Paper award from the Military Operations Research Society (MORS). And Jain et al. (2010) won the 2011 David Rist Prize from MORS for developing software tools that enable a similar randomization strategy to be used by LAX airport police when scheduling canine patrols and mobile checkpoints.

However, Cox (2008a) criticizes this decision rule on the grounds that it does not accurately reflect strategic thinking. We do not defend the plausibility of this rule, but it is conveniently simple for an introductory illustration. Chapter 2 considers alternatives.

Since Daphne is not telepathic, it is unlikely that she knows Apollo's true utilities. So her second step is to describe her uncertainty about those utilities through a subjective probability distribution. Specifically, suppose Daphne believes that:

- a successful smallpox attack provides Apollo a utility U_1 that is uniformly distributed on $[8, 10]$;
- a failed smallpox attempt provides Apollo a utility U_1^* that is uniformly distributed on $[-6, -2]$;
- a successful anthrax attack provides Apollo a utility U_2 that is uniformly distributed on $[4, 8]$;
- a failed anthrax attack provides Apollo a utility U_2^* that is uniformly distributed on $[-3, -1]$.

For this toy problem, simple distributions were chosen for the random utilities, to make the computation transparent. In practice, Daphne might have detailed insight into those utilities, based on public statements made by Apollo or a psychological profile created by her staff.

Daphne's third step is to assess Apollo's beliefs about his chances for success. Let p_1 and p_2 denote the probabilities that Apollo thinks are his chances for a successful smallpox or anthrax attack, respectively. Daphne is unlikely to know these: to her, they are random variables P_1 and P_2, and she uses subjective distributions to express her uncertainty. For simplicity, suppose that Daphne believes that P_1 is uniformly distributed on $[0.4, 0.7]$ and P_2 is uniformly distributed on $[0.6, 0.9]$.

With these assumptions, Daphne can now calculate her belief about Apollo's expected utility from mounting an attack with smallpox or anthrax. For a smallpox attack, she believes his expected utility is $X_1 = P_1 U_1 + (1 - P_1) U_1^*$. The random variables P_1, U_1 and U_1^* are independent, so her best guess of Apollo's expected utility from a smallpox attack is $\mathbb{E}[X_1] = 9(0.55) - 4(0.45) = 3.15$. Similarly, for anthrax, she finds $\mathbb{E}[X_2] = 4$.

Daphne's model for Apollo's decision process is that he will choose his attack randomly with probabilities proportional to his expected utility. Under that model, Daphne's ARA implies that she believes Apollo will choose smallpox with probability $\mathbb{E}[X_1]/(\mathbb{E}[X_1] + \mathbb{E}[X_2]) = 3.15/(3.15 + 4) = 0.44$ and choose anthrax with probability $4/(3.15 + 4) = 0.56$.

Now that Daphne has derived her best subjective belief $\pi_D(a)$ about the attack probabilities, she can apply straightforward decision analysis. For each of her choices, and for each of Apollo's attacks and their outcomes, she knows her utility $u_D(a, d, s)$. She also knows $p_D(s \mid a, d)$, her personal probabilities that Apollo's attack is successful, conditional on both her and Apollo's choices. Using (1.2), she chooses to stockpile smallpox vaccine or Cipro according to which one maximizes her expected utility.

In order to complete this example with a specific calculation, consider the values in Table 1.3. These show Daphne's losses (the negative of her utilities) under all possible outcomes. In our scenario, Daphne can only realize a loss, since even with a failed attack, she has incurred the stockpiling cost. The table indicates that the smallpox vaccine is more expensive than Cipro, and that Daphne's loss from a successful smallpox attack is greater than her loss from a successful anthrax attack.

Table 1.3 The cells show the loss (or negative utility) that Daphne believes she will suffer under all possible combinations of bioterrorist attack, defense, and success or failure of the attack.

| | Daphne Stockpiles Vaccine (d_1) | | Daphne Stockpiles Cipro (d_2) | |
	Apollo succeeds	Apollo fails	Apollo succeeds	Apollo fails
Smallpox attack (a_1)	20	2	150	1
Anthrax attack (a_2)	100	2	15	1

In order to calculate her expected loss for these utilities, Daphne must use her personal assessments of the probabilities that Apollo can stage a successful attack with smallpox and anthrax. Obviously, these probabilities may be very different from the ones she thinks Apollo holds about his chances of success. Suppose Daphne thinks Apollo has 1 chance in 100 of mounting a successful smallpox attack, and 1 chance in 20 of mounting a successful anthrax attack. Then her expected utility from decision d_1, stockpiling smallpox vaccine, is

$$\mathbb{E}[u_D(d_1,A,S)] = \sum_{s\in\mathcal{S}}\sum_{a\in\mathcal{A}} u_D(d_1,a,s)p_D(s\,|\,d_1,a)\pi_D(a)$$

$$= (-20)(0.01)(0.44) + (-100)(0.05)(0.56)$$

$$+ (-2)(0.99)(0.44) + (-2)(0.95)(0.56),$$

which comes to -4.8232. Note that in this simple example, $p_D(s\,|\,d_1,a)$ does not actually depend upon d_1, since her model for Apollo is that he does not attempt to analyze her decision making when selecting his attack. Similarly, for the decision regarding Cipro, Daphne's expected utility from d_2 is

$$\mathbb{E}[u_D(d_2,A,S)] = (-150)(0.01)(0.44) + (-15)(0.05)(0.56)$$

$$+ (-1)(0.99)(0.44) + (-1)(0.95)(0.56),$$

which is -2.0476. With these (almost arbitrary) numbers, Daphne maximizes her expected utility by choosing to stockpile Cipro.

This toy example is, of course, unrealistic. Daphne's assessment of Apollo's probabilities is simplistic; the supporting analysis for the utilities is absent. Also, the random probabilities P_1 and P_2 are independent, but since both are related to Apollo's resources and skills, there should be some association. And, to expedite the example, the question of how Daphne self-elicits all of these numbers was ignored. Chapter 2 discusses more plausible implementations of strategic analysis.

In sequential games, ARA is useful to the first-mover, who can optimize her choice by modeling the utilities, probabilities and capabilities of the second-mover. But ARA is not relevant to the second-mover, since he can observe the action chosen by the first-mover. To show how ARA works to support the first-mover, we revisit Example 1.3.

Example 1.3, continued: Daphne must decide whether to install body scanners at airport checkpoints, and Apollo must decide whether to attempt to smuggle a bomb onto an airplane. Daphne makes her choice first, which is observed by Apollo. We assume Daphne knows her own utility function and can specify her personal probabilities for the possible outcomes conditional on Apollo's choice. But she also needs to have a distribution over the choice that Apollo will make. ARA asserts that Daphne should build a model for Apollo's decision making, using distributions to express her uncertainty about Apollo's utilities and probabilities. This process imposes a distribution over Apollo's choice, and Daphne then selects the action that maximizes her expected utility.

As before, there are many ways in which Daphne might model Apollo's thinking. For example, she might study his previous public statements, in order to assess his utility from a successful attack and also infer what he believes are his chances of success. Or she might examine historical terrorist attempts, and assume that Apollo's probabilities and utilities are like those of his predecessors. If she believes that Apollo is rational, she thinks he will act so as to maximize his expected utility; if she believes he is irrational, then her model must reflect his defect, perhaps based on one of several theories of bounded rationality. If she is unsure whether he is rational, then she will assign probabilities to both possibilities and use a mixture model to describe his reasoning.

To illustrate the kind of calculation Daphne must make, suppose her model for Apollo's decision making is that he tosses an unfair coin, and attempts to smuggle a bomb on board if it comes up heads. She thinks that if Apollo observes that Daphne has installed body scanning equipment, then the coin has probability 0.1 of coming up heads; if he sees that she has not installed the equipment, the coin has probability 0.8 of coming up heads. This model is naive—it posits a puerile mechanism for Apollo's decision making that avoids strategic dependence of his choice upon his probabilities and utilities, as will be developed in Chapter 3. But for now, it satisfices the expository purpose.

Having built this model for Apollo's rule, Daphne must now examine her own utilities and probabilities. She knows that the cost of installing new body-scanning technology is 2 units. Experimental testing shows the new millimeter wave body scanning equipment improves bomb detection probability from 0.75 to 0.95, compared to the current X-ray scanner. Moreover, she believes her cost if Apollo successfully smuggles the bomb on board is 20 units. And she has no additional loss, beyond the installation cost, if Apollo's attempt is discovered and he is arrested, or if he is deterred from making the attempt. Table 1.4 shows the losses associated with every possible outcome.

Table 1.4 The cells show the loss (negative utility) that Daphne believes she will suffer when she installs or does not install the new scanning equipment, Apollo either attempts to smuggle a bomb or does not, and, if he attempts to smuggle, Apollo is either successful or not. "NA" indicates an impossible outcome.

	Daphne Installs Scanners (d_1)		Daphne Does Not Install Scanners (d_2)	
	Apollo succeeds	Apollo fails	Apollo succeeds	Apollo fails
Attempt	20+2	2	20	0
No attempt	NA	2	NA	0

Daphne now uses these numbers and her beliefs about the relevant probabilities to calculate the expected utility for each of her two possible choices. The expected utility from installing the detectors, choice d_1, is

$$\mathbb{E}[u_D(d_1,A,S)] = \sum_{s \in S} \sum_{a \in A} u_D(d_1,a,s) p_D(s \mid d_1,a) \pi_D(a \mid d_1)$$
$$= (-22)(0.1)(0.05) + (-2)(0.1)(0.95) + (-2)(0.9) = -2.1.$$

The calculation simplifies in the case when Apollo is deterred from attempting to smuggle the bomb, since the chances of success or failure are irrelevant. Similarly, for d_2, Daphne's expected utility is $(-20)(0.8)(0.25) = -4$. Thus Daphne maximizes her expected utility by installing the new millimeter wave scanners.

The examples considered in this chapter show the importance of the information structure in ARA. One must know what is private knowledge, and what is common knowledge. The influence diagram is a tool that helps analysts keep track of the information structure in complex games.

1.3 Influence Diagrams

An influence diagram (ID) is a graphical tool used to represent a decision problem. It is a directed acyclic graph with three kinds of nodes: decision nodes, shown as rectangles; chance nodes, shown as ovals; and preference nodes, shown as hexagons. (As a matter of terminology, sometimes the chance node is called an uncertainty node, and the preference node may be called a value node.) The ID generalizes Bayesian networks (cf. Pearl, 2005), and is an alternative to decision trees that avoids the proliferation of branches in complex problems, while making explicit the probabilistic and functional dependencies.

Each node in an ID has a domain that consists of the possible values that the variable corresponding to that node can take. Thus, the domain of a decision node is the set of possible choices for the decision maker. The domain of a chance node is the set of values that the corresponding random variable can take. And the domain of a preference node is the set of possible utilities, one of which the decision maker will obtain.

Arrows, or directed edges, between nodes describe the structure of the problem. An arrow that points to a chance node signifies that the distribution at that node is conditioned on the values of all nodes at its tail. An arrow that points to a preference node means that the utility function depends upon the values of all nodes at its tail. And an arrow that points to a decision node means that the choice made at that node is selected with knowledge of the values of all nodes at its tail.

As a simple example, consider the ID in Fig. 1.2. It describes a problem in which Daphne, the decision maker, is trying to decide whether an approaching hurricane should cause her to order the evacuation of New Orleans. The National Weather Service has reported that the hurricane may strike the city, and it provides probabilistic information regarding that event. Additionally, Daphne has knowledge (or beliefs) about the costs and benefits of ordering an evacuation, under every possible scenario regarding the strength of the hurricane.

In Fig. 1.2, the oval labeled "Hurricane" is a chance node—its values indicate how dangerous the hurricane will actually be, ranging from the storm completely missing the city to a catastrophic Category 5 landfall at high tide. The second oval, labeled "Forecast," is also a chance node. Its values are the probabilities supplied by the National Weather Service. The arrow from "Hurricane" pointing to "Forecast" indicates that the impact of this hurricane can be predicted, to some degree, from the meteorological measurements. The arrow from "Forecast" to the decision node labeled "Evacuate?" indicates that Daphne will know the weather forecast at the time she makes the decision on whether to evacuate. The lack of an arrow from

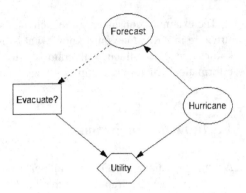

Fig. 1.2 The influence diagram for a decision on whether to evacuate before a hurricane.

"Hurricane" to "Evacuate?" indicates that Daphne will not know the hurricane's actual strength or landfall location at the time she must decide. Her decision node takes only two values: evacuate New Orleans, or do not evacuate. The arrows from "Evacuate?" and "Hurricane" to the preference node "Utility" indicate that the costs and benefits of both possible decisions must be evaluated under every possible outcome for the hurricane.

Daphne can use this ID as a tool for selecting the decision that maximizes her expected utility. To do that, she first needs the following inputs:

- At the preference node, she needs to know her utilities $u(e, h)$ as a function of both her decision on whether to evacuate (e) and how damaging the hurricane actually is (h).
- At the chance node "Hurricane" she needs to know her prior belief $p(h)$ about the probability distribution of the random damage H from the hurricane.
- At the chance node "Forecast" she needs to know her beliefs about the conditional distribution of the forecast F given that the damage from the hurricane is H, or $p(f \mid h)$. This will enable her to reverse the conditioning, and find her belief about the true strength of the hurricane given the forecast. Her belief $p(f \mid h)$ captures what she thinks about the accuracy of the forecaster.

These inputs come from Daphne herself, based on her experience with the costs of previous evacuations and previous hurricanes, and the reliability of previous forecasts. There is a large literature on how to elicit these quantified subjective beliefs (cf. O'Hagan et al., 2006).

Once the ID has been fully specified, Daphne can use backwards induction to solve it, eliminating nodes in the informational order defined by the ID. In this case the informational order is $F \prec E \prec H$. This ordering reflects the fact that the "Forecast" is known by Daphne when she must make her decision at the "Evacuate?" node and therefore comes first in the informational order, while the value of the "Hurricane" node is unknown and therefore comes last. As a consequence, the order in which the nodes of the ID are eliminated by the backwards induction algorithm is then $H \rightarrow E \rightarrow F$.

1. In order to eliminate the chance node "Hurricane," Daphne must first reverse the conditioning between H and F using Bayes rule. She obtains her posterior on H after she learns the forecast $F = f$:

$$p(h \mid f) = \frac{p(f \mid h)p(h)}{\int p(f \mid h)p(h)\,dh}.$$

2. Daphne can now eliminate the chance node "Hurricane" by integrating the utility function $u(e, h)$ with respect to the conditional $p(h \mid f)$ to find the expected utility of the two possible evacuation decisions given that she knows $F = f$:

$$\psi(e, f) = \int u(e, h)\, p(h \mid f)\,dh.$$

3. Finally, Daphne can eliminate the decision node "Evacuate?" by selecting the decision with maximum expected utility conditional on the forecast:

$$e^*(f) = \text{argmax}_e\, \psi(e, f).$$

In this way, Daphne can use the ID as a blueprint for decision making, given the available information and her personal utilities and probabilities. This blueprint becomes especially valuable in more complex problems.

IDs provide a high-level view of decision problems. They show the dependence among variables as well as the state of information at the time decisions are made, but they do not show the possible values associated with each decision or chance variable. That kind of detailed information is usually referenced in associated tables; see Clemen and Reilly (2004) for examples.

In ARA, the focus is upon problems with more than one decision maker. Such cases require an extension of IDs to Multi-Agent Influence Diagrams (MAIDs). MAIDs were proposed in Koller and Milch (2003), which proved that they give equivalent solutions to game trees, provide Nash equilibrium solutions when the minimax perspective is used, and can be solved by divide-and-conquer algorithms. From the ARA perspective, MAIDs are attractive because they provide a visualization that clearly distinguishes private knowledge from common knowledge.

As with IDs, MAIDs use rectangles to indicate decisions, ovals to indicate probability distributions, and hexagons to indicate preferences. Each decision node is owned by one of the agents, as shown by color or a label; the owner controls the value of that node. Similarly, each decision maker has a preference node, also distinguished by a color or label.

MAIDs may be viewed as the superposition of several individual IDs, where the ID of each decision maker interacts with one or more IDs of the other agents. Thus a MAID consists of IDs for each agent, with sufficient arrows between nodes to create a connected directed acyclic graph. Typically, there are separate decision and preference nodes for each decision maker, but the chance nodes may be shared among decision makers. When chance nodes are shared, they include probability distributions for each of the agents who share it, representing their different beliefs about that random value.

The computation of traditional game theory solutions assumes that the utilities and beliefs of each agent for the preference and chance nodes of the MAID are common knowledge. For example, it is assumed that the agents' beliefs about a chance event are common knowledge even though they may have different beliefs. The ARA approach will relax this assumption and solve the problem for only one of the agents. In particular, ARA realistically assumes that each agent will only know his own beliefs and preferences and that these are not known to the others.

For example, consider the MAID in Fig. 1.3, which corresponds to the Colonel Blotto game in Example 1.1. The rectangles $\boxed{B}$ and $\boxed{K}$ represent the battalion allocations chosen by Colonel Blotto and Colonel Klink, respectively. The absence of arrows between these two nodes means that each makes his allocation without knowing the other's choice (as must happen in a simultaneous game). Payoffs are represented by hexagons. Since this is a zero-sum game, then $u_B = -u_K$ and one can use a single node to represent the outcome. The arrows from each colonel's decision node to the preference node indicate that the payoff depends upon both allocation decisions. Recall that the game is deterministic; therefore, no uncertainty nodes are used in this graphical representation. Also, note that the detailed information on allocations and outcomes is not explicitly shown; one must refer to Table 1.1.

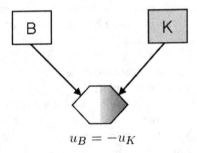

Fig. 1.3 The MAID for the deterministic Blotto game in Example 1.1.

$$u_B = -u_K$$

Figure 1.4 shows the representation of the generalization of the Blotto game described in the example from Sect. 1.2 in which the outcomes from tied allocations are stochastic. To show this, the MAID now has a chance node S that describes the randomness in the outcomes at the three battlefields. Also, note that this diagram allows for non-zero-sum outcomes. Here, each colonel has his own preference node, indicating that they may value a victory or defeat in different ways.

The ARA approach advises one of the decision makers against the others (in contrast to classical game theory, which finds an equilibrium solution for all the players simultaneously). When analyzing a decision problem from the perspective of a specific decision maker, one builds models for the behavior of the other actors, allowing the supported decision maker to maximize his expected utility.

To show how ARA can support Colonel Blotto's analysis, consider the ID shown in Fig. 1.5 representing the analysis described by Table 1.2. This is a restriction of the previous MAID in Fig. 1.4 that represents the problem just from Colonel Blotto's

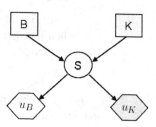

Fig. 1.4 The MAID for the Blotto game with uncertain outcomes/payoffs and different utility functions for the two colonels.

perspective. The ID deletes Colonel Klink's preference node (since it is irrelevant to Colonel Blotto) and converts Colonel Klink's decision node into a chance node. As a comparison, Fig. 1.5 shows the corresponding decision tree for Colonel Blotto's analysis.

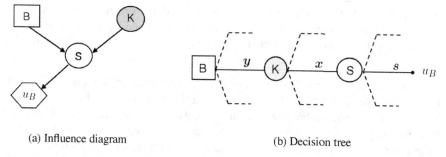

(a) Influence diagram

(b) Decision tree

Fig. 1.5 Two views of Colonel Blotto's decision analysis.

The graphical structure of the ID must satisfy the following properties:

- There are no arrows from value nodes to decision or chance nodes. If there is more than one value node, then each represents a different component of a multi-attribute utility function. (A multi-attribute utility function considers the payoff to be multivariate; e.g., Colonel Blotto may value the same outcome in different and incommensurate ways; e.g., according to the number of lives lost, the cost in matériel, the influence on the war, and his chance for promotion.) Sometimes a value node representing aggregated utility values is also included in the ID. In that case arrows are drawn from the value node for each attribute to the aggregated value node (which has no successors). These arrows represent the functional dependence of the multi-attribute utility which combines the utilities of different components of the outcome into a single quantity to be maximized.
- There is a directed path, containing all decision nodes, which specifies the order in which decisions are made by the decision maker: $D_1 \rightarrow D_2 \rightarrow \cdots \rightarrow D_m$, for m decision nodes. This implies the existence of a time or precedence-based ordering and induces a partition of the set of chance

nodes $\mathcal{C}$ into the subset $\mathcal{C}_0$ consisting of those random events that are known when the first decision D_1 is made; and the sets $\mathcal{C}_i$ consisting of those chance nodes whose values are observed between D_{i-1} and D_i, for $i = 1, \ldots, m-1$; and the set $\mathcal{C}_m$ of all chance nodes that are not observed, and therefore are unknown, before any of the decisions is made. Some of the sets $\mathcal{C}_i$ may be empty. This partition defines a partial order over decision and chance nodes: $\mathcal{C}_0 \prec D_1 \prec \mathcal{C}_1 \prec \cdots \prec \mathcal{C}_{m-1} \prec D_m \prec \mathcal{C}_m$. That partition determines the information structure given by the temporal order of decisions, specifying what information is known at the time each decision is taken. This structure makes it possible to convert IDs into decision trees.

- It is reasonable to assume that the decision maker recalls all previous decisions and observations. Thus any information available at a previous decision node is also available at all subsequent decision nodes, and so there should be an informational arrow from any node in $\bigcup_{j<i} D_j \cup \mathcal{C}_j$ into decision node D_i for all $i = 1, \ldots, m$. By convention, only the arrows whose origins are chance nodes in $\mathcal{C}_{i-1}$ or the decision node D_{i-1} are shown in an ID. The other arrows, called *no-forgetting arcs*, are not explicitly represented in the ID.

These properties imply that not every decision problem can be represented as an ID. If there is no predetermined order in which decisions and observations occur, then an ID is not appropriate. For example, in medical diagnosis or project troubleshooting, there may be many possible tests and no clearly specified order in which the tests should be performed. Further, the choice of subsequent tests depends upon the results of previous tests. In these cases, an (asymmetric) decision tree is a better graphical tool, with different subtrees indicating each of the possible test orders. But, as usual, the decision tree representation quickly generates too many branches. As an alternative, one might solve multiple IDs, one for each possible test order, and compare the solutions.

Often, the context of an ARA problem determines the order in which decisions are made, possibly with stages at which several decision makers must choose simultaneously. This happened in Example 1.3, where Apollo chooses a bioterrorist agent (smallpox or anthrax) for his attack and Daphne defends by stockpiling vaccine or Cipro. Figure 1.6 shows the MAID for this problem.

It is a simultaneous game, and the MAID shows this by the absence of an arrow between the decision nodes $\boxed{D}$ and $\boxed{A}$. The game has an asymmetric information structure: the net costs to Daphne from stockpiling vaccine or Cipro, which is shown in node $\widehat{c_D}$ whose domain is just the two costs, are known only to her. Similarly, the net costs to Apollo from mounting a smallpox or anthrax attack, which is shown in node $\widehat{b_A}$ whose domain is just those two expenses, are known only to Apollo. Both nodes are chance nodes, since Daphne can only have probabilistic beliefs about Apollo's expenses, and Apollo has only probabilistic knowledge about Daphne's costs. Daphne's private information about her costs c_D is represented by an arrow from $\widehat{c_D}$ to $\boxed{D}$. The arrow indicates that these costs are known by her at the time she makes her decision. The lack of an arrow from c_D to A means that Apollo does not know Daphne's costs at the time he makes his decision. The analogous rule applies to Apollo's private information. Figure 1.6 shows that the payoffs to each player depend upon both players' actions as well as upon the private information each has. For Daphne, it is shown by the three arrows into her preference node: one from her decision node, one from Apollo's decision node, and one from her private information on costs. Similarly, Apollo's preference node has three corresponding arrows.

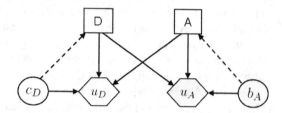

Fig. 1.6 The MAID for Example 1.3, concerning bioterrorism, showing the role of each agent's private information.

Figure 1.6 describes a game in which the outcome is deterministic, given the actions of each opponent. In practice, there may be an element of chance—perhaps an informant betrays the attack, or the vaccine is ineffectual. To capture this issue, Fig. 1.7 adds a chance node S. As before, the MAID consists of two coupled IDs, one for Daphne and the other for Apollo, but now with a shared uncertainty node.

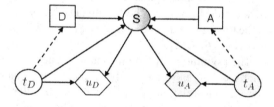

Fig. 1.7 The MAID for Example 1.3 on bioterrorism, with uncertain attack outcome and players' private information represented by types.

Following the MAID conventions for classical game theory, one represents the private information in terms of Harsanyi's theory of types (see Section 1.2). To Apollo, Daphne's type is unknown. She may be a type who has large costs for vaccine or Cipro, or she may be a type who has small costs; she may believe that Apollo has a large chance of success, or a small one; and she may weigh human life lightly or heavily. Similarly, to Daphne, Apollo's type is unknown. He may be able to produce smallpox easily, or not; he may be optimistic about his chance of success or not; and he may value success more or less highly. But Daphne knows her own type, and this is shown by the dashed arrow from her type t_D to her decision node D, and similarly for Apollo.

The arrows from Daphne's type node t_D to both her preference node and to the shared chance node S indicate that her utilities and probabilities are private information which depend on her type. Similarly, the arrows from Apollo's type

node show how his utilities and probabilities affect the decision, but are unknown
to Daphne. The implication is that Apollo has only probabilistic beliefs about the
utilities that will determine Daphne's decision, and Daphne has only probabilistic
beliefs about the utilities that will determine Apollo's decision.

The Bayes Nash equilibrium solution to this game assumes there is a common
prior distribution $\pi(t_D, t_A)$ over the types, so this distribution is mutually known,
and known to be known. Thus Daphne's beliefs about t_A, or $\pi_D(t_A) = \pi(t_A \mid t_D)$,
and Apollo's beliefs about t_D, or $\pi_A(t_D) = \pi(t_D \mid t_A)$, are known to the other player,
respectively. This assumption is necessary for each player to be able to solve simultaneously the game from the other's perspective. The computation of a Nash
equilibrium always presumes that players in the game have common knowledge of
each other's beliefs and preferences. The Harsanyi model is not an exception. Private
information about, e.g., Daphne's preferences is represented by a commonly known
utility function that is dependent on Daphne's type, which is known to her but not to
Apollo. Since Daphne's type is unknown to Apollo, his analysis requires him to form
beliefs about her type, and these beliefs are assumed to be common knowledge.

But a common prior is a strong and unreasonable assumption. Why would a player
disclose his distribution, and if he did, why would his opponent believe he was sin-
cere? In ARA, the problem is analyzed from the perspective of only one of the decision makers, so the analyst needs only beliefs about the other player's type in order
to maximize expected utility and solve the game. The ARA approach converts the
MAID into an ID, giving a precedence-based order over the decision and chance
nodes. For Daphne's ID this order is $t_D \prec D \prec \{t_A, A, S\}$, where A is now a chance
node from Daphne's perspective, and it provides her a blueprint for the solution. This
partial ordering reflects the fact that Daphne's type is the only variable known to her
at the time she makes her decision. Similarly, from Apollo's perspective, the order is
$t_A \prec A \prec \{t_D, D, S\}$.

The asymmetry of the information structure in this game makes it impossible to
jointly define an order for the decisions and chance nodes. To use a MAID, it is
necessary to relax the properties of the IDs that lead to sequencing and information structure. To ensure consistency between the informational structure and the
ordering of the decision makers' analyses, we require that if any two decisions are
simultaneous, then there is no directed path between them. It follows that a MAID
is an acyclic directed graph over the decision, chance and utility nodes of each decision maker, where chance nodes can be shared, under the constraint that, from each
decision maker's perspective, the MAID is a proper ID.

To check whether a MAID satisfies this constraint, one must test whether the
diagram generated from the MAID for each player is a proper ID. For each player,
an ID is obtained by deleting the other decision makers' preference nodes in the
MAID and converting their decision nodes into chance nodes. The chance nodes that
are not shared with the chosen player are eliminated. Also, decision nodes owned
by other decision makers may become barren nodes and thus be eliminated. Each
player's ID must then define a total order of decisions and a corresponding partial
order of chance nodes.

In this way, MAIDs can generate IDs that provide useful guidance in the context of ARA. MAIDS are used throughout this book to help the reader visualize complex information structures.

Exercises

1.1. Solve the hurricane example in Section 1.3. Assume that the National Weather Service forecasts that with probability f the hurricane will strike the city (and that Daphne believes it is accurate). Her utility associated with whether or not the city is evacuated and whether or not the hurricane strikes is as follows:

	Hurricane strikes ($H = 1$)	Hurricane does not strike ($H = 0$)
City evacuated ($E = 1$)	1	w
City not evacuated ($E = 0$)	0	1

where $0 < w < 1$. For what values of (w, f) should Daphne order an evacuation?

1.2. In the sequential Defend-Attack game in Example 1.3, Daphne receives utility 1 if Apollo does not smuggle a bomb onto the airplane and utility 0 if he does. The cost (negative utility) to Daphne for installing millimeter wave body scanners is c. Give the solution as a function of Daphne's probabilities $t_0 = p_D(A = 1 \mid d = 0)$ and $t_1 = p_D(A = 1 \mid d = 1)$, where $A = 1$ denotes a smuggling attempt, and she has probabilities $v_1 = p_D(S = 1 \mid d = 1, a = 1)$ and $v_0 = p_D(S = 1 \mid d = 0, a = 1)$ for the chance that smuggling is successful, with and without body scanners, respectively.

1.3. For Exercise 1.2, help Daphne assess t_0 and t_1 by analyzing Apollo's decision problem. Suppose an informant reliably reports that Apollo's utility function is:

A	S	$u_A(a,s)$
1	1	1
1	0	0
0		x

$A = 1$ means that Apollo attempts to smuggle a bomb, $A = 0$ means he does not, $S = 1$ means he is successful, and $S = 0$ means he is not. Assume Daphne believes $x \sim \text{Unif}(0, 0.5)$. The informant also reports that Apollo thinks body scanning technology halves his chance $p_A(S = 1 \mid d = 0, a = 1)$ of successful smuggling. Daphne's subjective belief is that $p_A(S = 1 \mid d = 0, a = 1) \sim \text{Unif}(0.1, 0.2)$. Find Daphne's t_0 and t_1, as well as the choice that maximizes her expected utility, when $v_0 = 0.2$, $v_1 = 0.1$ and $c = 0.01$.

1.4. Find the two pure strategy Nash equilibria for the *Game of Chicken*. This is a two-person simultaneous game in which drivers race toward each other, and either swerve or go straight at the last minute. The payoff bimatrix is:

	Swerve	Straight
Swerve	T , T	L , W
Straight	W , L	C , C

where W (win) $\succ$ T (tie) $\succ$ L (lose) $\succ$ C (crash). Each player prefers the other driver to be the one to swerve, but if both decide to drive straight they will crash, which is the worst possible outcome.

1.5. *Matching Pennies* is a zero-sum game in which two players, each with a penny, choose simultaneously which face to display. If both faces agree (both heads or both tails), then the row-player wins. If they disagree, the column player wins. The payoff matrix for this game (from the row player's perspective) is

	Heads	Tails
Heads	1	−1
Tails	−1	1

Find a Nash equilibrium for this game.

1.6. Suppose one modified the Game of Chicken and the Matching Pennies game described in Exercises 1.4 and 1.5 so that in both cases the row player's choice was made first, after which the column player must choose (knowing the row player's choice). Would this advantage or disadvantage the column player?

1.7. Suppose $n > 2$ individuals in a group are each given \$10 and the opportunity to invest the money in a common project whose benefits will be equally distributed among themselves. Specifically, each member of the group can put up to \$10 into a *common* pot that gets doubled and then split evenly among the group members, regardless of their individual contributions to the pot. Suppose also that these contributions are kept private: participants will not know how much the others have contributed. What amount should the group contribute to the common pot in order to maximize their joint benefit? Find the Nash equilibrium in this situation and compare it with the ideal group contribution.

1.8. Find the Nash equilibria for the two simultaneous games determined by the following payoff matrices.

	Game I	
	Change	Status Quo
Change	10 , 10	−1 , 0
Status Quo	0 , −1	0 , 0

	Game II	
	Change	Status Quo
Change	10 , 10	−1 , 0
Status Quo	0 , −∞	0 , 0

Intuitively, would one expect the same outcome from these two games? Justify possible differences in the results from pairs of choices in these games from both the game-theoretic and the decision-analytic perspectives.

Chapter 2
Simultaneous Games

This chapter describes adversarial risk analysis (ARA) for a two-person discrete simultaneous game. Specifically, it considers ARA for cases when the opponent is non-strategic, or employs one of several representative solution concepts: the Nash equilibrium, level-k thinking, and a mirroring equilibrium.

The main goals of ARA are to weaken the standard common knowledge assumptions used by Nash equilibrium solution concepts and to provide more flexible models for opponent behavior. ARA is explicitly Bayesian, in that subjective distributions are employed to express the uncertainties of the analyst. In particular, ARA distinguishes three different kinds of uncertainty:

- *Aleatory uncertainty* concerns the randomness of outcomes conditional on the choices made by both opponents.
- *Epistemic uncertainty* concerns the strategic choices of an intelligent adversary, as driven by unknown preferences, beliefs, and capabilities.
- *Concept uncertainty* concerns beliefs about how the opponent frames the problem. What kind of strategic analysis is used? Is the opponent rational? How deeply does he think? Often, the solution concept that is used will determine the epistemic uncertainties that are relevant.

Each of these categories may require subsidiary modeling, in much the same way that risk analysis for a bridge failure may combine uncertainties about construction methods, geology, and traffic load. In a comparison of several methods for risk management, Merrick and Parnell (2011) prefer ARA precisely because it handles these separate uncertainties more explicitly.

In most situations, after the opponents both choose their actions, the outcome for each is a random variable. In the context of counterterrorism, Apollo may choose to bomb a train, while Daphne has policemen search the train. In that case, the payoff is a random variable—there is a chance that the policemen will thwart the attack, and a chance that the bomb will explode, killing a random number of people and causing a random amount of economic and political damage. This randomness is aleatory uncertainty. It conditions on the choices of the opponents and, therefore, does not depend upon any strategic calculation on their parts.

In contrast, epistemic uncertainty describes Daphne's distribution over the choice Apollo will make, which usually integrates his preferences, beliefs, and capabilities. For the bomb-on-a-train example, Daphne does not know which train Apollo will target. His choice depends upon which train is his most valuable target (a preference), which train he thinks he has the best chance of attacking (a belief), and whether or not he has a bomb (a capability). Daphne does not know these, and thus has epistemic uncertainty which she expresses as a distribution over all possible trains. Similarly, Apollo does not know how many policemen Daphne will assign to a given train, and he can express his epistemic uncertainty through a subjective probability distribution.

Concept uncertainty arises from ignorance of how one's opponent will frame the analysis. In the language of classical game theory, this means that Daphne does not know which solution concept Apollo will use to make his decision. If Daphne thinks that Apollo is rational and that he is using a Nash equilibrium solution concept for a zero-sum game, then she expects him to maximize his minimum gain for the circumstance in which Daphne selects the action least favorable to him. But if she thinks Apollo is not "rational," then she would have to model his decision making through, say, prospect theory (cf. Kahneman and Tversky, 1979). And if she thinks he is non-strategic, then she believes he will select an attack without any consideration given to the choices that Daphne might make. Concept uncertainty embraces a wide range of strategies, and is an essential component in the ARA formulation.

Bayesian techniques enable great flexibility in tailoring the analysis to the situation, and in accounting for the different kinds of uncertainty that arise. Often the problem involves a hierarchy of decision analyses (in chess, these levels are called *plies*, which indicate how many moves ahead someone thinks). Such hierarchies are similar to the level-k models used in game theory (cf. Stahl and Wilson, 1994, 1995), but the ARA formulation is fully Bayesian and its implementation is closer to risk analysis than game theory. The following discussion explores different aspects of the ARA framework in the context of discrete simultaneous games.

2.1 Discrete Simultaneous Games: The Basics

A discrete two-person simultaneous game (or *normal form* game) between, say, Daphne and Apollo is naturally represented as an $m \times n$ bimatrix X with entries (X_{ij}^D, X_{ij}^A), where X_{ij}^D and X_{ij}^A are, respectively, the payoffs to Daphne and Apollo, when he makes choice j and she makes choice i. The rows of the bimatrix correspond to the possible actions of Daphne; the columns correspond to the possible actions of Apollo. (When there are $r > 2$ players, the bimatrix representation generalizes to an r-dimensional array.)

In most practical situations, the payoffs in the cells are not fixed numbers but rather random variables. The two opponents often have different beliefs about the distributions of those random variables, and imperfect knowledge of what each other believes. For example, in a simultaneous Defend-Attack game, Apollo may believe that he has a good chance of success and a large payoff, whereas Daphne thinks his attack will probably fail. Neither knows the specific probabilities the opponent assigns, nor does either know the specific utilities of their opponent. Such situations violate the framework used in traditional game theory, especially the common knowledge assumption needed to implement the Bayes Nash equilibrium solution and related concepts.

To a decision analyst, the bimatrix formulation is helpful because it distinguishes epistemic uncertainty (i.e., which row–column pair is chosen, given the selection of a specific solution concept) from aleatory uncertainty (the outcome from picking that row–column pair). Within any specific cell determined by the row–column choice, Daphne can apply traditional probabilistic risk analysis methods based upon expert opinion, probability models, historical data, and so forth, as described in, e.g., Bedford and Cooke (2001) or Cox (2013). Her analysis generates a distribution over the result when she and Apollo choose that row–column pair of actions. By combining that distribution with her own utility function, she can calculate the distribution for her payoff. A similar analysis allows her to infer the distribution that Apollo has for his payoff, and this enables deeper reasoning related to epistemic uncertainty.

All interesting features of a bimatrix game are present in the simultaneous Defend-Attack example. Daphne, the Defender, chooses from a finite set of actions $D = \{d_1, \ldots, d_m\}$. Apollo, the Attacker, chooses from the finite set $A = \{a_1, \ldots, a_n\}$. To simplify notation, the following discussion suppresses the subscripts: (d, a) represents any specific (d_i, a_j). For each pair of choices (d, a), Daphne receives the utility $u_D(d, a, \omega)$ which depends upon both chosen actions and upon chance, as indicated by the random variable ω. (In problems with fixed, non-random payoffs, one omits the ω.) Daphne's belief about the probability distribution for ω, conditional on the choice (d, a), is represented by $p_D(\omega \mid d, a)$. Symmetrically, Apollo receives the utility $u_A(d, a, \omega)$, and believes the conditional density of ω is $p_A(\omega \mid d, a)$. Then Daphne's expected utility, given the choices (d, a), is

$$\mathbb{E}[u_D(d, a, \omega) \mid d, a] = \int u_D(d, a, \omega) p_D(\omega \mid d, a) \, d\omega.$$

Similarly, Apollo's expected utility is $\int u_A(d, a, \omega) p_A(\omega \mid d, a) \, d\omega$.

Although one can work abstractly with utilities, from a modeling perspective it is simpler to first find the distributions of outcomes conditional on a specific pair of actions (d, a), and then find the corresponding utilities. For example, Daphne could use the probability model $p_D(\omega \mid d, a)$ to describe her belief about the chance of not discovering a bomb, where d is Daphne's allocation of policemen to trains and a is Apollo's decision about which train to target. Then, conditional on the outcome that the bomb is not discovered, Daphne can separately assess her utility, which combines mortality, economic costs, and political capital. Since it is cognitively challenging to develop the probability model, and also cognitively challenging to assess complex utilities, it is better not to conflate these tasks. The notation indicates this separation by writing Daphne's random outcome as $Y_D(d, a, \omega)$ and her realized utility as $u_D[Y_D(d, a, \omega)]$. Referring back to the bimatrix $\boldsymbol{X}$, the entry (X_{ij}^D, X_{ij}^A) indicates that

$X_{ij}^D = u_D[Y_D(d_i, a_j, \omega)]$ and $X_{ij}^A = u_A[Y_A(d_i, a_j, \omega)]$. Here $Y_A(d_i, a_j, \omega)$ is the random outcome for Apollo, who has utility function $u_A(\cdot)$. Figure 2.1 visualizes these relationships in the form of a MAID.

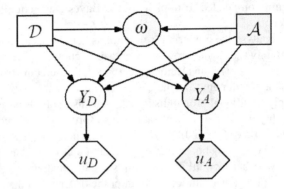

Fig. 2.1 A Multi-agent influence diagram showing the decision, chance, and utility nodes, together with the shared information structure, for the simultaneous Defend-Attack problem.

To perform an ARA, the first step is to address the concept uncertainty. The second one handles epistemic uncertainty, and the third step deals with aleatory uncertainty. Each step carries its own challenges. The following discussion describes an idealized analysis that treats each of these uncertainties in a fully Bayesian way. Since the idealized analysis is often too complex for practical implementation, there are also helpful heuristics that allow workable approximations.

Regarding concept uncertainty, Daphne must model how Apollo will make his decision. But there are many possible solution concepts that he might use. These include random strategies, boundedly rational strategies, and so forth. Some common solution concepts are:

1. *Non-strategic play*, in which Daphne believes that Apollo will select an action without consideration of her choice. This includes the case in which Apollo selects actions with probability proportional to the perceived utility of success (cf. Paté-Cornell and Guikema, 2002); it also includes non-sentient opponents, such as a hurricane.
2. *Nash equilibrium or Bayes Nash equilibrium methods*, both of which imply that Daphne believes Apollo is assuming that he and Daphne have a great deal of common knowledge.
3. *Level-k thinking*, in which Daphne believes Apollo thinks k plies deep in an "I think that she thinks that I think ..." kind of reasoning. The level-0 case corresponds to non-strategic play.
4. *Mirroring equilibrium analysis*, in which Daphne believes Apollo is modeling Daphne's decision making in the same way that she is modeling his, and both use subjective distributions on all unknown quantities.

Usually, Daphne does not know which solution concept Apollo has chosen. But based on previous experience with Apollo, and perhaps input from informants or other sources, she can place a subjective probability distribution over his possible solution concepts. She could then make the decision that maximizes her expected utility against that weighted mixture of strategies.

Realistically, a full Bayesian analysis that puts positive probability on a large number of different solution concepts becomes computationally burdensome. But in principle, the approach is simple. Each solution concept will lead (after handling the relevant epistemic and aleatory uncertainties) to a distribution over Apollo's actions. Then Daphne weights each distribution by her personal probability that Apollo is using that solution concept. This generates a weighted distribution on $\mathcal{A}$, Apollo's action space, which reflects all of Daphne's knowledge about the problem and all of her uncertainty. The approach is closely related to Bayesian model averaging, in which uncertainty about the model is expressed by a probability distribution over the possible models, and inference is based upon a weighted average of the posterior distributions from each model (cf. Clyde and George, 2004; Hoeting et al., 1999).

Regarding epistemic uncertainty, this is handled differently for each solution concept that Daphne thinks Apollo might use. The kinds of epistemic uncertainty statements that are needed depend upon the solution concept. For example, with the Nash equilibrium concept, Daphne believes that Apollo thinks they both know the same bimatrix of payoffs. In that case, the relevant epistemic uncertainty is Daphne's distribution over the bimatrices that Apollo may be using. But for the Bayes Nash equilibrium concept, there is additional uncertainty related to Harsanyi's theory of types (Harsanyi, 1967a). Besides the distribution over the bimatrices, Daphne must also express her epistemic uncertainty about the common knowledge distributions which Apollo is assuming that they share (i.e., what he thinks are their common distributions on types). In principle, full description of the epistemic uncertainties is complex, even with simple solution concepts. Often, there are pragmatic approximations that may be used. Subsequent examples in this chapter illustrate these issues.

Regarding aleatory uncertainty, this concerns the non-strategic randomness in an outcome. Given a particular row–column choice in the bimatrix, the payoffs to each party are usually stochastic. In that case, Daphne must assess her beliefs about the outcome probabilities, conditional on the row–column pair. It warrants emphasis that this is not the same as assessing her beliefs about what Apollo's distributions over those outcomes might be—that is a matter of epistemic uncertainty, since it requires her to model Apollo's reasoning.

Aleatory uncertainty can be addressed through traditional probabilistic risk analysis (Bedford and Cooke, 2001). Daphne's beliefs should be informed by expert judgment, previous history, and appropriate elicitation methods (cf. O'Hagan et al., 2006). Regrettably, risk analysis may be imprecise: experts are overconfident, previous history may be only partially relevant or even misleading, and the wide range of elicitation methods highlights the pitfalls in making complex judgments. Nonetheless, probabilistic risk analysis is a mature field and the following examples provide insight into how it may be used in the course of performing an ARA.

2.2 Modeling Opponents

The distinguishing feature of ARA is that it emphasizes the advantage of building
a model for the strategic reasoning of an opponent. As a practical matter, one does
not play chess in the same way against a novice as one would play against a grand-
master. Instead, one makes a judgment about the skill and style of the opponent, and
plays accordingly. In sequential games, that judgment is updated as the game unfolds,
but at each decision point the amount of effort and the acceptable risk depend upon
how strong the opponent seems to be. This is in contrast with the situation in classical
game theory, where each player assumes that all other players are as smart as John
von Neumann.

   In particular, in many real-world cases, such as terrorism, the opponent may be
neither clever nor rational. Some terrorists hold extreme political or religious views
that cannot be squared with reason. Some terrorists work in groups, in which case
their actions are determined by complex social dynamics and internal politics rather
than by pure rationality. And some terrorists are so opportunistic that their actions
are more like random weather events than strategic choices. As an example of pre-
dictable irrationality, note that many recent attacks have been timed to occur on national holidays or the anniversaries of previous attacks. This pattern provides coun-
terterrorists an advantage; e.g., airports often arrange for heightened security on those
days.

   The remainder of this section develops ARA solutions in the context of four common models for an opponent's reasoning: non-strategic, Nash equilibrium, level-k,
and mirroring. The subsection on level-k thinking also includes a comparison with
the Bayes Nash equilibrium solution. Most examples are treated in the context of a
semi-realistic toy problem, in order to concretely illustrate how the decision maker
may use domain knowledge and subjective beliefs to frame the analysis.

2.2.1 Non-Strategic Analysis

The simplest non-strategic game is one against a non-sentient opponent. In that case,
traditional risk analysis has long been accepted as the appropriate approach. For example, it is widely used in planning for natural disasters, for policy making, and
for large classes of business decisions (Singpurwalla, 2010). In such probabilistic
risk analyses, the decision maker has a distribution over the kinds of events that
may occur, and distributions over the costs of actions to mitigate or remedy the consequences. All of these distributions reflect aleatory uncertainty (e.g., What is the
probability that a Category 5 hurricane will strike New Orleans? How many people
would die? What would it cost to evacuate the city?)

   Most decision analysts would agree that one should select the action that maximizes expected utility (cf. French and Ríos Insua, 2000). However, there is a little
bit of disagreement in the case of low-probability, high-consequence situations, especially when there is substantial uncertainty (cf. Samuelson, 1977; Haimes, 2004).

The following examples simply maximize expected utility, but see the discussion in later chapters.

Example 2.1: Captain Hornblower must sail from Mumbai to Rome. But Somali pirates are a threat to international shipping (Ploch et al., 2009). Until recently, naval warships did not patrol the Horn of Africa, and it was illegal for ships to travel with armed private security, so the ship captain had two choices. He could make a short but risky voyage through the Suez Canal, or undertake a longer but safer journey around the Cape of Good Hope. In his analysis, Captain Hornblower assumes that the pirates are not strategic planners, which seems reasonable given that they have few options other than to roam the sea in search of prey.

Using historical data on piracy off the coast of Somalia, Carney (2009) estimates that a ship is attacked with probability 0.005, and, conditional on an attack, that it is successfully hijacked with probability 0.4. He also estimates, from data on six recent hijacks, that the average ransom paid is €2.3M, which this analysis takes as the expected value of the random ransom. Also, assume that if the ship is not hijacked when attempting the Suez Canal, then the additional cost to its owners is €0. However, if Captain Hornblower decides to sail around the Cape of Good Hope, the trip will cost an additional €0.5M. Table 2.1 summarizes these costs.

Table 2.1 Expected costs of the routes under all possible scenarios.

Route	Attack, Hijacked	Attack, Not Hijacked	Not Attacked
Suez Canal	2.3	0	0
Cape of Good Hope	N/A	N/A	0.5

Figure 2.2 represents the information in the table through a decision tree.

Captain Hornblower's decision depends upon his utility function for money. There are several special cases, including:

- His utility function is $u_1(x) = x$, in which case he is *risk neutral*.
- His utility function has the form $u_2(x) = 1 - \exp(-\alpha x)$, in which case he has *constant absolute risk aversion* (CARA).
- His utility function has the form $u_3(x) = (x - \alpha)^{1-\beta}/(1 - \beta)$, in which case he has *hyperbolic absolute risk aversion* (HARA).

This terminology derives from the Arrow-Pratt measure of absolute risk aversion, defined as $A(x) = u''(x)/u'(x)$, where $u(x)$ is the utility function (Pratt, 1964). The measure has useful mathematical properties, but from a behavioral standpoint, the important fact is that most people tend to be risk averse (Holt and Laury, 2002). Risk-averse people have concave utility functions, such as all the ones described

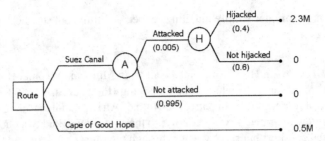

Fig. 2.2 A decision tree that shows the possible outcomes from Captain Hornblower's decision, with their probabilities and payoffs.

above (u_1 is a special case—it is both concave and convex, and hence it is risk neutral). A risk-averse person prefers a small guaranteed payoff to a random payoff that has larger expected value but some chance of being very small. Concave utility is the foundation of the insurance industry.

Note that in Captain Hornblower's problem, consequences are expressed in costs, rather than benefits, and therefore the analysis employs $u(-x)$, where x is the corresponding cost in the decision tree. Captain Hornblower must weigh the uncertain cost of the Suez passage against the certain cost from rounding the Cape of Good Hope.

Table 2.2 shows Captain Hornblower's decision analysis under a range of utility functions, where x is in millions of euros. For the Cape of Good Hope route, the expected utility is $u(-0.5)$, where $u(\cdot)$ is one of the utility functions. For the Suez Canal transit, the expected utility is

$$0.005 \times 0.4 \times \mathbb{E}_F[u(-X)] + 0.005 \times 0.6 \times u(0) + 0.995 \times u(0),$$

where F is the distribution of the ransom that is paid for a captured vessel; to produce the table, it is assumed that F is uniform between €2M and €2.6M.

Table 2.2 Expected utility for the two routes under utility functions in the Arrow-Pratt family.

Utility Function	Expected Utility	
	Suez Canal	Cape of Good Hope
Risk Neutral, u_1	**−0.005**	−0.500
CARA, $\alpha = 0.5$	**−0.004**	−0.284
CARA, $\alpha = 2$	**−0.211**	−1.719
CARA, $\alpha = 4$	−24.929	**−6.389**
HARA, $\alpha = -3, \beta = 0.25$	**3.035**	2.651
HARA, $\alpha = -3, \beta = 0.5$	**3.461**	3.162

Table 2.2 shows that Captain Hornblower's optimal decision depends upon his utility function. If he is risk neutral, or uses CARA with small risk-aversion coefficients, he should transit via the Suez Canal, as he would do if he uses the HARA utility functions proposed. For larger CARA coefficients, he should transit around the Cape of Good Hope.

This toy calculation used fairly conventional utility functions. But people are more complex than that. For example, empirical work on the endowment effect shows that most people count losses more heavily than gains (Kahneman, Knetsch and Thaler, 1990).

2.2.2 Nash Equilibrium

The minimax principle is the simplest example of the Nash equilibrium solution concept. It applies to zero-sum games, where both players have common knowledge of the payoff matrix. From a practical standpoint, most games are not precisely zero-sum. And even if one has a zero-sum game, rarely do both players know their common utilities. Nonetheless, this solution concept is a reasonable approximation in some circumstances.

For example, consider the toy problem in Example 1.2, for which Apollo will either develop an anthrax attack or a smallpox attack, and Daphne will stockpile either Cipro or smallpox vaccine. Neither party has the capability to do both. In that situation, Apollo might reason that his development expenses and Daphne's defense investments are both sunk costs—their respective budgets for offense and defense have been allocated, the full amounts will be expended, and thus those costs are irrelevant to their strategic calculation. In that case, Apollo could rationalize that this bioterrorism game is zero-sum, with payoffs equal to the number of lives that are lost or saved.

If Daphne believes that Apollo has framed the problem as a zero-sum game, then she believes that Apollo will choose his attack so as to maximize the minimum number of deaths (a maximin approach; equivalently, he is minimizing the maximum number of survivors). She may believe this because Apollo has used the maximin solution concept previously, or perhaps an informant has tipped her off, or perhaps this is just one component of Daphne's comprehensive analysis, in which she puts probability weights on several different solution concepts that Apollo might use. Whatever the basis, the Nash equilibrium solution concept requires that Daphne address her epistemic uncertainty about the values in Apollo's payoff matrix.

Let the payoff matrix for Apollo be

	Smallpox	Anthrax
Vaccine	W	Y
Cipro	X	Z

where Daphne's payoffs are implicitly represented as the negative of Apollo's pay-offs, since this is a zero-sum game. If Daphne knew Apollo's (W, X, Y, Z) values, she could apply the maximin principle to solve the game and discover the action Apollo would choose, enabling her to make the best response.

Typically, Daphne will not know Apollo's payoff values. The ARA perspective suggests she should treat these as random variables. Daphne's joint distribution over the (W, X, Y, Z) values that Apollo uses might be based upon medical knowledge of the pathogens, military intelligence from informants, personal intuition, or all of these and more. Eliciting joint probability distributions that combine information from multiple sources is non-trivial, but for now, assume Daphne has done so, and that her personal probability about the payoffs in Apollo's table is represented by the joint density function $f(w, x, y, z)$.

In this situation, some analysts find the expected values of W, X, Y, and Z, then substitute these into the payoff matrix and solve the corresponding game. But this approach is logically flawed—the operations of finding the minimax solution and taking expectations do not commute. The value of the minimax solution to the expected game is not the expected value of the minimax solution.

Instead, Daphne needs to calculate her personal probability p^* that Apollo will attack with smallpox, with $1 - p^*$ the chance of an anthrax attack. This p^* is the weighted average of the probabilities of an smallpox attack over all possible values of (w, x, y, z), or

$$p^* = \int_{-\infty}^{\infty} \int_{-\infty}^{\infty} \int_{-\infty}^{\infty} \int_{-\infty}^{\infty} \mathbb{P}[\text{ smallpox attack } | w, x, y, z] f(w, x, y, z) \, dw \, dx \, dy \, dz.$$

To solve this integral, Daphne must consider the 24 possible ways in which W, X, Y, and Z might be ordered, from least to largest. These correspond to 24 disjoint regions of $\mathbb{R}^4$, denoted by $R_1, \ldots, R_{24}$. (For brevity and clarity, this analysis ignores the possibility of tied values—for continuous joint densities, ties occur with probability zero, so this simplification would not affect the conclusion.)

A region contains a *saddlepoint* if a value in the payoff table is simultaneously a row minimum and a column maximum. When the table has one of them, then Daphne should choose that row and Apollo should choose that column; neither can unilaterally improve their outcome by making a different choice. In this two-by-two game, there can be at most one saddlepoint. The row–column pair for W is a saddlepoint if $X > W > Y$. The other possible saddlepoints are X, when $Z > X > W$; or Y, when $W > Y > Z$; or Z, when $Y > Z > X$. For the case when W or X is a saddlepoint, the minimax theorem implies that Daphne knows that Apollo is certain to attack with smallpox. When Y or Z are saddlepoints, she knows he will attack with anthrax.

The saddlepoint solution is a pure strategy, in the sense that each player will always make a fixed choice. In contrast, when the table has no saddlepoint, then the minimax solution is a mixed strategy: each player selects an action at random, with probabilities determined by the values in the table. In general, the mixed strategies are found by linear programming, but for a two-by-two example, there is a closed form expression. Specifically, suppose $W > Z > X > Y$. This game has no saddlepoint solution, and simple algebra shows that Apollo maximizes his expected gain when he chooses the smallpox attack with probability $p = (Z-X)/(W+Z-X-Y)$; see, e.g., Rapoport (1970, Chap. 7). There are seven other orderings that do not have a saddlepoint solution, and rearrangement of the formula finds the smallpox attack probabilities for each case.

By considering the pure and mixed strategy solutions for each of the 24 regions of integration, Daphne can calculate p^*, her personal probability that Apollo will attack with smallpox, taking account of all her uncertainties about the payoff table Apollo has assumed they are both using. Given that, her decision is simple.

Daphne can use self-elicitation to find her beliefs about the expected number of lives lost under each possible pair of choices (i, j), where i indicates her choice and j indicates Apollo's. Denote these expected values by μ_{ij}, as below:

	Smallpox	Anthrax
Vaccine	μ_{11}	μ_{12}
Cipro	μ_{21}	μ_{22}

Daphne's expected loss from stockpiling vaccine is $p^*\mu_{11} + (1-p^*)\mu_{12}$ and her expected loss from stockpiling Cipro is $p^*\mu_{21} + (1-p^*)\mu_{22}$. She calculates, and then selects the action that minimizes the expected number of deaths.

2.2.3 Level-k Thinking

A level-k analysis allows one to model how deeply an opponent reasons about a game (cf. Stahl and Wilson, 1994, 1995). If Daphne performs a level-1 analysis, she assumes that Apollo is a level-0 thinker; i.e., his choice is non-strategic, and depends only upon his own payoffs or perhaps is made at random. A level-2 analysis means that Daphne believes that Apollo is a level-1 thinker, who will model Daphne as a level-0 thinker. A level-3 analysis means that Daphne believes that Apollo is a level-2 thinker, and so forth. In this framework, Daphne wants to reason one level deeper than Apollo.

As an example, consider the two payoff matrices below:

	Left	Right
Up	0, ?	10, ?
Down	10, ?	0, ?

	Left	Right
Up	0, 0	10, 10
Down	10, 0	0, 10

In the left matrix, Daphne sees only her own payoffs, and neither Up nor Down is a dominant choice for her. Since she does not know Apollo's payoffs, and since she is not using ARA methods to place a subjective probability distribution over those payoffs, then she cannot model his reasoning. She is forced to be a level-0 thinker. The most common decision rules used in these situations are:

- The minimax criterion, in which one minimizes the largest possible loss. This is equivalent to the maximin rule, which maximizes the smallest possible gain.
- The minimax regret criterion, in which one minimizes the maximum difference between the payoff that was realized and the payoff that was possible in hindsight. (This criterion is not as pessimistic as the minimax criterion.)
- The Hurwicz criterion, in which one maximizes the weighted average of the best and worst payoffs associated to each alternative. The weight $\alpha \in [0,1]$ given to the best payoff from each choice is called the optimism coefficient. When $\alpha = 0$, the Hurwicz criterion is equivalent to the minimax rule.
- The Laplace criterion, in which one maximizes the average payoff. This puts equal weight upon all of Apollo's possible moves, implying that Apollo's choices are equiprobable.

For the left-hand matrix, none of these approaches can produce a clear recommendation and Daphne must therefore choose arbitrarily.

Now suppose that she knows the full bimatrix on the right, which also contains Apollo's payoffs. Daphne can apply level-1 thinking to see that Right is Apollo's dominant choice, and thus her best play is Up. Daphne is using *iterated dominance*, in which she successively eliminates dominated choices in order to select her best move (cf. Myerson, 1991, Chap. 2.5). Iterative dominance assumes players are level-k thinkers.

In level-k analyses, ARA offers a natural way to account for the epistemic uncertainty about unknown payoffs, thus moving beyond the traditional Nash equilibrium formulation. For example, consider the following bimatrix, where the value of (X, Y) is unknown to Daphne: she believes it has probability π of being $(-4, 3)$ and probability $1 - \pi$ of being $(3, -4)$.

	War	Peace
Fight	X, Y	$1, -1$
Friend	$-1, 2$	$0, 0$

This bimatrix describes a game of conflict. If both players choose reject conflict, then both receive a payoff of 0. A unilateral attack by Daphne has payoff 1 to her and loss 1 to Apollo; a unilateral attack by Apollo has payoff 2 to him and loss 1 to Daphne. When both attack, there is payoff 3 for the stronger opponent and loss 4 for the weaker one.

If the outcome of a battle were decided by chance, then the uncertainty is aleatory. But assume here that the stronger opponent prevails and that Apollo knows whether he is stronger or weaker than Daphne; i.e., Apollo knows the value of (X,Y). Then Daphne's uncertainty is epistemic, since it pertains to Apollo's private knowledge. This asymmetric knowledge mimics a terrorist threat; for instance, terrorists know whether or not they have the capability to deploy weaponized smallpox, but the United States probably does not.

To begin, suppose Daphne performs a level-1 analysis, in which she models Apollo as a non-strategic thinker. If $(X,Y) = (-4,3)$, then Apollo is stronger and his dominant choice is War. So (if he is rational) he would play War, and Daphne believes this will occur with probability π. But if $(X,Y) = (3,-4)$, then Apollo is weaker and has no dominating strategy. In that case, Daphne must assign a personal probability p to the event that Apollo chooses War, and $1 - p$ to the event that he chooses Peace. (Her guess about this p might be based upon previous experience, informants, or even simple intuition.) So Daphne's expected utilities are

$$\begin{aligned}
\psi_D(\text{ Fight }) &= -4 \times \mathbb{P}[\text{ War and (-4,3) }] + 3 \times \mathbb{P}[\text{ War and (3,-4) }] + 1 \times \mathbb{P}[\text{ Peace }] \\
&= -4\pi + 3p(1-\pi) + (1-p)(1-\pi) \\
\psi_D(\text{ Friend }) &= -1 \times \mathbb{P}[\text{ War }] + 0 \times \mathbb{P}[\text{ Peace }] \\
&= -\pi - p(1-\pi).
\end{aligned} \tag{2.1}$$

Algebra shows Daphne should choose Fight if and only if $p > (4\pi - 1)/(3 - 3\pi)$.

Next suppose that Daphne does a level-2 analysis, which assumes that Apollo is a level-1 thinker. So Daphne thinks Apollo assumes that she is a level-0 thinker. Since Daphne has no dominant choice, Apollo must place a probability over her selection. Let q be his probability that she will choose Fight. Then Apollo calculates his expected utilities as follows. When he is stronger than Daphne, then

$$\begin{aligned}
\psi_A(\text{ War }) &= 3 \times \mathbb{P}[\text{ Fight }] + 2 \times \mathbb{P}[\text{ Friend }] = 3q + 2(1-q) = q + 2 \\
\psi_A(\text{Peace}) &= -1 \times \mathbb{P}[\text{ Fight }] + 0 \times \mathbb{P}[\text{ Friend }] = -q
\end{aligned}$$

and he would always choose War. When he is weaker than Daphne, then

$$\begin{aligned}
\psi_A(\text{ War }) &= -4 \times \mathbb{P}[\text{ Fight }] + 2 \times \mathbb{P}[\text{ Friend }] = -4q + 2(1-q) = 2 - 6q \\
\psi_A(\text{Peace}) &= -1 \times \mathbb{P}[\text{ Fight }] + 0 \times \mathbb{P}[\text{ Friend }] = -q
\end{aligned}$$

and he should choose War if and only if $q \leq \frac{2}{5}$.

To complete Daphne's level-2 analysis, she places a subjective distribution F over the value of q that Apollo ascribes to her. When Apollo is stronger, she thinks he must surely choose War, which happens with probability π. When he is weaker, the probability that he will choose War is $p = \mathbb{P}[q \leq \frac{2}{5}] = F(\frac{2}{5})$. Solving (2.1), her expected utilities are

$$\begin{aligned}
\psi_D(\text{ Fight }) &= -4\pi + 3F(\tfrac{2}{5})(1-\pi) + (1 - F(\tfrac{2}{5}))(1-\pi) \\
\psi_D(\text{ Friend }) &= -\pi - F(\tfrac{2}{5})(1-\pi).
\end{aligned}$$

Daphne should choose Fight if and only if $F(\frac{2}{5}) > (4\pi - 1)/(3 - 3\pi)$.

It is slightly tedious, but one can go further. Suppose Daphne thinks Apollo is more subtle than before. She does a level-3 analysis, which assumes Apollo is a level-2 thinker. Thus Apollo thinks Daphne is a level-1 thinker, who models him as a level-0 thinker.

In this framework, Daphne thinks that a level-0 Apollo will always attack when he is stronger, which she believes happens with probability π, and when he is weaker, she believes he attacks with probability p. As a level-1 thinker, Daphne's best action is to attack when $p > (4\pi - 1)/(3 - 3\pi)$. When a level-2 Apollo is weaker he will calculate his expected utilities as

$$\psi_A(\text{ War }) = -4 \times \mathbb{P}[\text{ Fight }] + 2 \times \mathbb{P}[\text{ Friend }] = -4q + 2(1 - q)$$
$$\psi_A(\text{Peace}) = -1 \times \mathbb{P}[\text{ Fight }] + 0 \times \mathbb{P}[\text{ Friend }] = -q,$$

where q is his probability that $p > (4\pi - 1)/(3 - 3\pi)$. Apollo will attack when $q = \mathbb{P}[p > (4\pi - 1)/(3 - 3\pi)] \leq \frac{2}{5}$.

Finally, as a level-3 thinker, Daphne will assess a subjective probability distribution over Apollo's q by eliciting a joint distribution $G(p^*, \pi^*)$, where

- π^* is what she believes Apollo thinks is her estimate of the chance that he is stronger, and
- p^* is what she believes is what he thinks is her probability that he will choose Left when he is weak. (He always chooses War when he is stronger.)
- G is a distribution, which itself is random, that describes Daphne's uncertainty about Apollo's beliefs (p^*, π^*) over Daphne's probabilities (p, π).

From the previous level-1 analysis, Daphne thinks Apollo thinks that she will choose Fight when $p^* > (4\pi^* - 1)/(3 - 3\pi^*)$. And from the previous level-2 analysis, Apollo should think Daphne will choose Fight with probability $Q = \mathbb{P}_G(\mathcal{S})$ where $\mathcal{S} = \{(p^*, \pi^*) : p^* > (4\pi^* - 1)/(3 - 3\pi^*)\}$. So, when Apollo is weak, Daphne thinks his calculation is

$$\psi_A(\text{ War }) = -4 \times \mathbb{P}[\text{ Fight }] + 2 \times \mathbb{P}[\text{ Friend }] = -4Q + 2(1 - Q)$$
$$\psi_A(\text{ Peace }) = -1 \times \mathbb{P}[\text{ Fight }] + 0 \times \mathbb{P}[\text{ Friend }] = -Q.$$

Thus, Daphne believes Apollo will choose War when $Q \leq \frac{2}{5}$.

Next, Daphne must calculate her own expected utilities based on the preceding analysis in order to decide whether to attack. Solving (2.1) gives

$$\psi_D(\text{ Fight }) = -4\pi + 3\mathbb{P}[Q \leq \frac{2}{5}](1 - \pi) + 1\mathbb{P}[Q > \frac{2}{5}](1 - \pi)$$
$$\psi_D(\text{ Friend }) = -\pi - \mathbb{P}[Q \leq \frac{2}{5}](1 - \pi).$$

Note that here she uses her true π, representing her best guess about the probability that Apollo is stronger. The π^* is a random variable representing what Apollo thinks is her best guess, elicited from Daphne's perspective through the random G.

If Daphne knows the joint distribution $G(p^*, \pi^*)$ used by Apollo, represented by a single non-random G, then $\mathbb{P}[Q \leq \frac{2}{5}]$ would be either 0 or 1. But it is more realistic to assume that she is uncertain about G. In that case she could place a Dirichlet process prior on the space of all distributions, then do an appropriate integral to find the distribution of $Q = \mathbb{P}_G(\mathcal{S})$ (Hjort et al., 2010). More simply, Daphne might specify a finite number of plausible distributions, say $G_1, \ldots, G_M$, and place subjective probability weights $w_1, \ldots, w_M$ upon them. In that case, Q would be a proper random variable with values $q_1, \ldots, q_M$ and probabilities $w_1, \ldots, w_M$, where $q_i = \mathbb{P}_{G_i}(\mathcal{S})$, and she could directly compute $\mathbb{P}[Q \leq \frac{2}{5}] = \sum w_i 1_E(i)$, where $1_E(i)$ is the indicator function for $E = \{i : q_i \leq \frac{2}{5}\}$.

The intricacy of this reasoning raises a key question: How large should k be in a level-k analysis? Lee and Wolpert (2012), Ho, Camerer and Weigelt (1998), and Stahl and Wilson (1995) provide experimental evidence that people do not usually go beyond level $k = 2$ or 3. There are different approaches to the problem of closing the potentially infinite regress, including:

1. The analyst picks a single k, based upon their belief about the opponent.
2. The analyst places a subjective distribution over k, solves the problem separately for each value of k, and then weights those solutions according to the subjective distribution.
3. The analyst places a noninformative prior at some reasonable level in the hierarchy, which closes the analysis.

The previous examples took the first approach, for $k = 1, 2, 3$. But the second approach is consistent with Bayesian principles, and should be preferred. The third approach, see e.g., Ríos Insua, Rios and Banks (2009), is a practical expedient (which could be used in conjunction with either of the first two analyses); it has the advantage of avoiding difficult elicitation of vague beliefs. Rothschild, McLay and Guikema (2012) arguably robustifies level-k solutions by placing a uniform distribution over Apollo's actions at the first rung of the ladder of level-k reasoning, and this could be combined with all three of these approaches.

Now it is possible to compare this Bayesian level-k analysis to the Bayes Nash Equilibrium (BNE) solution. Recall that the BNE analysis uses the theory of types and assumes that there is common knowledge over the distribution of player types, where each type has its own personal probabilities and utilities (epistemic uncertainty). Each player knows his own type, but not those of others. This framework converts a simultaneous game of incomplete information (about probabilities and utilities) into one of imperfect information (random types), in which Nature moves first, selecting the type for each player according to the mutually known prior distribution.

In the preceding example, Apollo had two possible types: with probability π, he is stronger than Daphne so $(X, Y) = (-4, 3)$; with probability $1 - \pi$ he is weaker, and $(X, Y) = (3, -4)$. Apollo knows his type, but Daphne does not. The critical BNE assumption is that π is known to both parties.

For the bimatrix in this example, one can show that:

- If $\pi > \frac{4}{7}$, then the only BNE pure strategy is (Friend, War); i.e., Apollo should attack and Daphne should stand down.
- If $\pi < \frac{1}{4}$, then the only BNE pure strategy is (Fight, σ), where σ is War if Apollo is stronger and Peace if Apollo is weaker.

These solutions are quite different from those found by any of the level-k analyses.

To confirm that these are BNE solutions, one must first prove that (Friend, War) is a Nash equilibrium when $\pi > \frac{4}{7}$. Apollo's best response against Friend is War, no matter whether his type is strong or weak. And Friend is Daphne's best response against War if and only if

$$u_D(\text{ Friend, War }) > \mathbb{E}[u_D(\text{ Fight, War })].$$

Since $u_D(\text{ Friend, War }) = -1$ and $\mathbb{E}[u_D(\text{ Fight, War })] = -4 \times \pi + 3 \times (1 - \pi)$, this is true when $\pi > \frac{4}{7}$.

Next, one shows that (Fight, σ) is a BNE when $\pi \leq \frac{2}{5}$. Daphne's expected utilities against Apollo's decision rule σ are:

$$\psi_D(\text{ Fight},\sigma) = -4 \times \mathbb{P}[\text{ Apollo is strong }] + 1 \times \mathbb{P}[\text{ Apollo is weak }]$$
$$= 1 - 5\pi$$
$$\psi_D(\text{ Friend},\sigma) = -1 \times \mathbb{P}[\text{ Apollo is strong }] + 0 \times \mathbb{P}[\text{ Apollo is weak }]$$
$$= -\pi.$$

So Fight is better for Daphne than Friend against σ whenever $\pi < \frac{1}{4}$.

Apollo's decision rule σ is his best response against Fight: If Apollo is stronger, then War is always better for him than Peace. If Apollo is weaker, then Peace is better than War when Daphne plays Fight.

2.2.4 Mirror Equilibria

Level-k thinking raises the possibility of a troublesome infinite regress in strategic calculation. An alternative solution concept is the *mirror equilibrium*, which avoids that regress by allowing the analyst to suppose that both opponents seek an equilibrium. However, since neither player knows the other's utilities and probabilities, the Bayes Nash equilibrium solution concept cannot apply. Instead, the analyst places a subjective distribution over the utilities and probabilities of the opponent.

The motivating example is convoy routing through a roadway network on which the opponent has placed improvised explosive devices (IEDs) at unknown locations. Dimitrov et al. (2011), Bayrak and Bailey (2008), and Washburn and Wood (1995) have studied similar routing problems using classical game theory. The ARA given here is based upon Wang and Banks (2011).

In this example, an IED does not prevent passage. But it causes a random amount of damage, and the independent random damage accumulates additively if multiple IEDs are encountered. It turns out that this independent and additive structure will ensure the existence of a unique equilibrium.

This analysis supports Daphne, the convoy commander, who wants to select a route that minimizes her expected total damage. Her opponent, Apollo, wants to place IEDs along the roadway network so as to maximize the damage Daphne receives. This comprises a normal form game, in that both the route selection and the IED locations are chosen in advance: Daphne does not alter her route in response to IED damage, and Apollo does not locate new IEDs on the fly, as Daphne's route choice is revealed.

Daphne must start at a location S in a fixed network, and choose a route that leads to a terminus T. These locations are known to Apollo, who may site IEDs at certain locations (the vertices, v_k) along the network. A simple example is shown in Fig. 2.3

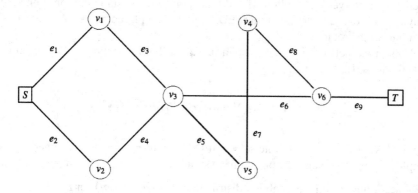

Fig. 2.3 An example of a road network.

Daphne's action space $\mathcal{D}$ consists of all routes connecting S to T (without backtracking). She has four possible choices:

$$d_1 : \ S \to v_1 \to v_3 \to v_5 \to v_4 \to v_6 \to T$$
$$d_2 : \ S \to v_1 \to v_3 \to v_6 \to T$$
$$d_3 : \ S \to v_2 \to v_3 \to v_5 \to v_4 \to v_6 \to T$$
$$d_4 : \ S \to v_2 \to v_3 \to v_6 \to T,$$

so $\mathcal{D} = \{d_1, d_2, d_3, d_4\}$. Daphne does not know how many IEDs Apollo has, nor does she know where he will place them. But she may have historical information on the numbers and locations of IEDs from previous convoy runs. Also, she may have intelligence from informants; such information could be unreliable, but it would nonetheless inform her subjective probability beliefs.

A pure strategy for Daphne is to deterministically select a specific route in $\mathcal{D}$. A mixed strategy for Daphne is to randomly select one of these routes according to some distribution $Q = (q_1, \cdots, q_J)$; i.e., route $d_j \in \mathcal{D}$ is chosen with probability q_j. Denote the set of all probability distributions over $\mathcal{D}$ as

$$\mathcal{Q} = \left\{ Q = (q_1, \cdots, q_J) : \sum_{j=1}^{J} q_j = 1 \text{ and } q_j \geq 0 \text{ for } j = 1, \ldots, J \right\}.$$

Since $\mathcal{Q}$ can put probability 1 on a particular route, this formulation includes both pure and mixed strategies.

Similarly, Apollo's pure strategies $\mathcal{A}$ are all the possible placements of his available IEDs. Usually, because of financial, time or human resource constraints, he cannot attack all of the vertices; the set $\mathcal{A}$ reflects this constraint. Each pure strategy $a \in \mathcal{A}$ may be represented by a binary column vector a of length K, where K is the number of vertices, with ones to indicate vertices at which IEDs are placed and zeroes to indicate vertices without IEDs. Innocuously, we assume that only one IED may be placed at a vertex, and that IEDS may not be placed at S or T, the start or terminus of the route.

In the road network example shown in Figure 2.3, if Apollo has chosen to use exactly three IEDs, then

$$\mathcal{A} = \left\{ a : a = (\alpha_1, \ldots, \alpha_K)' : \text{ for } \alpha_k = 0 \text{ or } 1 \text{ and } \sum_{k=1}^{K} \alpha_k = 3 \right\}.$$

Apollo uses a pure strategy if he selects IED locations non-randomly; otherwise he employs a mixed strategy. Apollo's mixed strategies result from choosing an element in $\mathcal{A}$ according to some probability distribution $P = (p_1, \cdots, p_I)$ for $I = \binom{K}{3}$; i.e., $a_i \in \mathcal{A}$ is chosen with probability p_i. Denote the set of all probability distributions over $\mathcal{A}$ as

$$\mathcal{P} = \left\{ P = (p_1, \cdots, p_I) : \sum_{i=1}^{I} p_i = 1 \text{ and } p_i \geq 0 \text{ for } i = 1, \ldots, I \right\}.$$

Since some elements of $\mathcal{P}$ put probability 1 on specific choices of locations, this formulation includes both pure and mixed strategies.

Besides the action spaces, one must also consider the payoffs to both adversaries. Daphne obtains a gain associated with the successful arrival of the convoy at T, but has random costs from damage in transit to the convoy. Also, there are costs specific to the route; e.g., long routes require more time and fuel. For Apollo, the gains are the value of inflicted damage and increased political capital, while the costs are the resources needed for the attack. Neither Daphne nor Apollo will have exact knowledge of the opponent's utility function, but military intelligence and previous experience enable them to make probability assessments of how the opponent values particular outcomes.

The analysis assumes that the utilities to both opponents are the additive sum of their incremental utilities across the entire route: the net cost to Daphne is the value of the mission less the sum of the damage from each IED encountered less the travel cost in fuel and time; the net gain to Apollo is the sum of the political success from the cumulative damage, less the cost of expended IEDs. It is necessary to assume that both Daphne's and Apollo's utilities can be expressed in commensurate units; e.g., they can both be monetized. (Monetization puts the value of human life, vehicle damage, and mission completion upon a common scale; it also puts political capital and expended IEDs on that same scale.)

Suppose that Daphne chooses route d_j which contains vertex k. Then her incremental utility when vertex k is encountered is the random variable Y_{jk}. Her random matrix of all such increments is denoted by Y, which is $J \times K$. Similarly, the incremental random utility for Apollo when Daphne selects route j containing vertex k is X_{jk}, and the random matrix of all such increments is X.

The vector a indicates Apollo's choice of IED sites. Let the vector d be a binary vector that indicates Daphne's choice of route. Choosing route d_j corresponds to the binary vector with 1 in position j and zeroes elsewhere. In this example $d_2 : S \rightarrow v_1 \rightarrow v_3 \rightarrow v_6 \rightarrow T$ corresponds to $d = (0,1,0,0)'$, and so forth. Then, the total payoff to Daphne if she chooses route d and Apollo chooses sites a is $d'Ya$, and the total payoff to Apollo is $d'Xa$.

This model assumes that the random damage done by an IED is independent of any previous IED damage. Thus the payoff from an IED at vertex k is independent of the path the convoy took in reaching vertex k, and so all random variables in the same column of the payoff matrix are equal. For the roadway network in Fig. 2.3, the payoff matrices are:

$$
Y = \begin{pmatrix}
 & v_1 & v_2 & v_3 & v_4 & v_5 & v_6 \\
\hline
d_1 & L_1 & 0 & L_3 & L_4 & L_5 & L_6 \\
d_2 & L_1 & 0 & L_3 & 0 & 0 & L_6 \\
d_3 & 0 & L_2 & L_3 & L_4 & L_5 & L_6 \\
d_4 & 0 & L_2 & L_3 & 0 & 0 & L_6
\end{pmatrix}, \quad
X = \begin{pmatrix}
 & v_1 & v_2 & v_3 & v_4 & v_5 & v_6 \\
\hline
d_1 & G_1 & 0 & G_3 & G_4 & G_5 & G_6 \\
d_2 & G_1 & 0 & G_3 & 0 & 0 & G_6 \\
d_3 & 0 & G_2 & G_3 & G_4 & G_5 & G_6 \\
d_4 & 0 & G_2 & G_3 & 0 & 0 & G_6
\end{pmatrix}. \quad (2.2)
$$

For clarity in the terminology, outcomes for Daphne are "losses" and outcomes for Apollo are "gains," and these are denoted by L_k and G_k, respectively. (There is no loss of generality since both losses and gains may be negative or positive.) Thus, if Daphne chooses route d_1, so that $d = (1,0,0,0)'$, and if Apollo places IEDs at v_2, v_3 and v_6, so that $a = (0,1,1,0,0,1)'$, then the net loss to Daphne is $d'Ya = L_3 + L_6$, whereas the net gain to Apollo is $d'Ya = G_3 + G_6$.

Suppose Daphne uses a mixed strategy $Q = (q_1, \ldots q_J)$, choosing route d_j with probability q_j. Similarly, suppose Apollo uses a mixed strategy $P = (p_1, \ldots, p_I)$, choosing the set of IED sites a_i with probability p_i. Then, the expected payoffs to Daphne and Apollo are, respectively,

$$\psi_D(Q,P) = \left(\sum_{j=1}^{J} q_j \boldsymbol{d}_j\right)' \boldsymbol{Y} \left(\sum_{i=1}^{I} p_i \boldsymbol{a}_i\right) \text{ and}$$

$$\psi_A(Q,P) = \left(\sum_{j=1}^{J} q_j \boldsymbol{d}_j\right)' \boldsymbol{X} \left(\sum_{i=1}^{I} p_i \boldsymbol{a}_i\right).$$

Now one can find the ARA mirroring equilibrium solution. Daphne must use her beliefs about the payoff functions and other available information to develop a subjective prediction for Apollo's behavior; i.e., her probability distribution over $\mathcal{A}$. Her distribution should reflect her assumption that Apollo is performing a similar analysis regarding her strategy. The term *mirroring* derives from this self-similar modeling of the opponent's decision making.

Daphne's goal is to find her probability distribution P over Apollo's strategy space $\mathcal{A}$; i.e., she believes that Apollo will place IEDs according to a_i with probability p_i. Finding P requires her to infer Q, the distribution she believes Apollo has over her choice of route in $\mathcal{D}$. Once Daphne has P, then, as an expected utility maximizer, she would select the route d^* (a pure strategy) that maximizes her expected utility

$$\psi_D(\boldsymbol{d},P) = \boldsymbol{d}'\boldsymbol{Y}\left(\sum_{i=1}^{I} p_i \boldsymbol{a}_i\right).$$

Thus, Daphne needs to derive P and Q, which she will do through a mirroring equilibrium argument. Let

- $\widetilde{\boldsymbol{X}}$ be the random matrix that describes Daphne's beliefs about Apollo's payoffs; and
- $\widetilde{\boldsymbol{Y}}$ be the random matrix that describes Daphne's beliefs about what Apollo believes is her payoff matrix.

These random variables are based upon a probability space $(\Omega, \mathcal{F}, \mathbb{P})$ that encodes all of Daphne's information, with generic outcome ω. For a given ω, the random payoff matrices are $\widetilde{\boldsymbol{X}}(\omega)$ and $\widetilde{\boldsymbol{Y}}(\omega)$, and the corresponding mixed strategies are $P(\omega)$ and $Q(\omega)$, respectively. (The formal probability space need not deter readers who are unfamiliar with measure theory. The key point is that Daphne has some distribution for what Apollo thinks will happen when he chooses a and she chooses d, and also a distribution for what he thinks Daphne thinks will happen in that circumstance.)

To begin, suppose that Daphne has a point-mass prior on $\omega \in \Omega$. This implies she believes that Apollo's payoff matrix is given by $\widetilde{\boldsymbol{X}}(\omega)$, and so Apollo would solve for the pure strategy

$$\underset{a \in \mathcal{A}}{\text{argmax}} \left(\sum_{j=1}^{J} q_j(\omega)\boldsymbol{d}_j\right)' \widetilde{\boldsymbol{X}}(\omega)\boldsymbol{a}. \tag{2.3}$$

Next, by allowing ω to have non-point-mass support, the probability distribution from $(\Omega, \mathcal{F}, \mathbb{P})$ imposes a distribution over (2.3), which gives a distribution on $\mathcal{P}$. In order to maximize her expected utility, Daphne must find the expected value of this distribution with respect to ω:

$$P = \mathbb{E}_{\mathbb{P}} \left[\underset{a \in \mathcal{A}}{\operatorname{argmax}} \left(\sum_{j=1}^{J} q_j(\omega) d_j \right)' \tilde{X}(\omega) a \right] \in \mathcal{P}.$$

This provides the probability distribution over $\mathcal{A}$, which is Daphne's prediction of Apollo's strategy.

On the other hand, knowing the state $\omega \in \Omega$ and given a prediction $P \in \mathcal{P}$ of Apollo's strategy, Daphne thinks that Apollo believes that she is trying to solve

$$\underset{Q \in \mathcal{Q}}{\operatorname{argmax}} \left(\sum q_j d_j \right)' \tilde{Y}(\omega) \left(\sum p_i a_i \right), \quad \forall \omega \in \Omega,$$

which yields a random vector $Q(\omega)$ on $(\Omega, \mathcal{F}, \mathbb{P})$ taking values in $\mathcal{Q}$.

These predicted probability distributions will be consistent if they satisfy the following definition:

Definition 2.1. The probability distribution $P^* \in \mathcal{P}$ and the family of probability distributions $\{\{Q^*(\omega) \in \mathcal{Q}\} : \omega \in \Omega\}$ constitute a *mirroring fixed point* given the analyst's information set $(\Omega, \mathcal{F}, \mathbb{P})$ if they simultaneously satisfy

$$P^* = \mathbb{E}_{\mathbb{P}} \left[\underset{a \in \mathcal{A}}{\operatorname{argmax}} \left(\sum_j q_j^*(\omega) d_j \right)' \tilde{X}(\omega) a \right],$$

$$Q^*(\omega) = \underset{Q \in \mathcal{Q}}{\operatorname{argmax}} \left(\sum q_j d_j \right)' \tilde{Y}(\omega) \left(\sum p_i^* a_i \right), \quad \omega \in \Omega. \tag{2.4}$$

A route choice d^{ARA} is said to be a pure-strategy ARA solution for Daphne if

$$d^{\text{ARA}} = \underset{d \in \mathcal{D}}{\operatorname{argmax}} d' Y \left(\sum_i p_i^* a_i \right), \tag{2.5}$$

where P^* is obtained as the fixed point in the mirroring analysis.

The mirroring fixed point is asymmetric in terms of how Daphne's information $(\Omega, \mathcal{F}, \mathbb{P})$ is used. Daphne's strategy is conditioned on her beliefs about what Apollo thinks about the problem, and what she believes Apollo thinks she believes about the problem. In contrast, the prediction for Apollo does not assume that he has access to Daphne's information.

It remains to show that the mirror equilibrium is a well-defined solution concept. To do this, one must prove the existence of the mirroring fixed point defined by (2.4). To that end, assume that $\Omega = \{\omega_1, \ldots, \omega_L\}$ is finite and let

$$\mathbb{P}[\omega_\ell] = \pi_\ell, \quad \ell = 1, 2, \ldots, L.$$

For any fixed pair of choices a and d, with vector representations $\boldsymbol{a}$ and $\boldsymbol{d}$, define the quantities

$$V_{a,d}^{(\ell)} := \boldsymbol{d}'\widetilde{X}(\omega_\ell)\boldsymbol{a}, \quad \ell = 1, 2, \ldots, L,$$

$$W_{a,d}^{(\ell)} := \boldsymbol{d}'\widetilde{Y}(\omega_\ell)\boldsymbol{a}, \quad \ell = 1, 2, \ldots, L,$$

$$\overline{W}_{a,d}^{(\ell,m)} := \pi_m W_{a,d}^{(l)}, \quad \ell, m = 1, 2, \ldots, L.$$

This notation enables the fixed point equations (2.4) for the probability distributions $P^* = (P_a^*)_{a \in \mathcal{A}} \in \mathcal{P}$ and $\left\{ Q^{*(\ell)} = (Q_d^{*(\ell)})_{d \in \mathcal{D}} \in \mathcal{Q} : \ell = 1, \ldots, L \right\}$ to be written as:

$$P^* = \sum_{\ell=1}^{L} \pi_\ell \left(\operatorname*{argmax}_{(P_a) \in \mathcal{P}} \sum_{a \in \mathcal{A}, d \in \mathcal{D}} P_a V_{a,d}^{(\ell)} Q_d^{*(\ell)} \right) \tag{2.6}$$

$$Q^{*(\ell)} = \operatorname*{argmax}_{Q_d \in \mathcal{Q}} \sum_{a \in \mathcal{A}, d \in \mathcal{D}} P_a^* W_{a,d}^{(\ell)} Q_d, \quad \ell = 1, \ldots, L. \tag{2.7}$$

Lemma 2.1. *If L probability distributions $\{P^{*(\ell)} \in \mathcal{P} : \ell = 1, 2, \ldots, L\}$ on $\mathcal{A}$ and L probability distributions $\{Q^{*(m)} \in \mathcal{Q} : m = 1, 2, \ldots, L\}$ on $\mathcal{D}$ satisfy the system of fixed-point equations*

$$P^{*(\ell)} = \operatorname*{argmax}_{(P_a) \in \mathcal{P}} \sum_{a \in \mathcal{A}, d \in \mathcal{D}} P_a V_{a,d}^{(\ell)} Q_d^{*(\ell)}, \quad \ell = 1, \ldots, L, \tag{2.8}$$

$$Q^{*(\ell)} = \operatorname*{argmax}_{(Q_d) \in \mathcal{Q}} \sum_{\substack{a \in \mathcal{A}, d \in \mathcal{D} \\ 1 \le m \le L}} P_a^{*(m)} \overline{W}_{a,d}^{(\ell,m)} Q_d, \quad \ell = 1, \ldots, L. \tag{2.9}$$

then $P^* = \sum_{\ell=1}^{L} \pi_\ell P^{*(\ell)}$ and $\left\{ Q^{*(\ell)} : \ell = 1, \ldots, L \right\}$ must satisfy (2.6) and (2.7) and therefore form a mirroring fixed point.

Proof: Multiply (2.8) by π_ℓ and sum over $\ell = 1, \ldots, L$ to recover (2.6). In (2.9), note that, by definition, for any $Q_d \in \mathcal{Q}$,

$$\sum_{m=1}^{L} P_a^{*(m)} \overline{W}_{a,d}^{(\ell,m)} Q_d = \sum_{m=1}^{L} P_a^{*(m)} \pi_m W_{a,d}^{(\ell)} Q_d = P_a^* W_{a,d}^{(\ell)} Q_d,$$

which yields (2.7). □

Lemma 2.2. *The fixed point for the system defined by (2.8) and (2.9) exists.*

Proof: For any ℓ and for any fixed $Q = (Q_d)_{d \in \mathcal{D}}$, the sum $\sum_{a \in \mathcal{A}, d \in \mathcal{D}} P_a V_{a,d}^{(\ell)} Q_d$ in (2.8) is linear and thus concave in the decision variables $P = (P_a)_{a \in \mathcal{A}}$. Also, for any ℓ and any fixed $\left\{ P^{(m)} = (P_a^{(m)})_{a \in \mathcal{A}} : m = 1, \ldots, L \right\}$, the sum

$$\sum_{\substack{a\in\mathcal{A},d\in\mathcal{D}\\ 1\leq m\leq L}} P_a^{(m)}\overline{W}_{a,d}^{(\ell,m)}Q_d$$

is also linear and thus concave in the decision variables $Q=(Q_d)_{d\in\mathcal{D}}$. Additionally, all feasible sets of decision variables are convex compact sets in a finite Euclidean space. So, for any $(\hat{P}_a^{(\ell)})_{a\in\mathcal{A}}\in\mathcal{P}$ and $(\check{P}_a^{(\ell)})_{a\in\mathcal{A}}\in\mathcal{P}$, one has

$$\sum_{a\in\mathcal{A}}\left[\gamma\hat{P}_a^{(\ell)}+(1-\gamma)\check{P}_a^{(\ell)}\right]=\gamma\underbrace{\sum_{a\in\mathcal{A}}\hat{P}_a^{(\ell)}}_{=1}+(1-\gamma)\underbrace{\sum_{a\in\mathcal{A}}\check{P}_a^{(\ell)}}_{=1}=1,\quad \ell=1,\dots,L$$

for all γ such that $0\leq\gamma\leq 1$, and hence $\gamma\hat{P}_a^{(\ell)}+(1-\gamma)\check{P}_a^{(\ell)}\in\mathcal{P}$. A similar argument applies to $\mathcal{Q}$. By the Nash Fixed-Point Theorem (cf. Aubin, 1993), a fixed point exists for the system defined by (2.8) and (2.9). $\square$

Combining Lemma 2.1 and Lemma 2.2 gives the key result.

Theorem 2.1. *The mirroring fixed point defined by* (2.4) *exists.*

Note that the fixed point to the system (2.8) and (2.9) may be regarded as the (mixed-strategy) Nash Equilibrium for a game in which there are L P-players (choosing $\{P^{(\ell)}\in\mathcal{P}:\ell=1,\dots,L\}$) and and L Q-players (choosing $\{Q^{(\ell)}\in\mathcal{Q}:\ell=1,\dots,L\}$) with payoff functions

$$\sum_{a\in\mathcal{A},d\in\mathcal{D}}P_a^{(\ell)}V_{a,d}^{(\ell)}Q_d^{(\ell)}\quad\text{and}\quad\sum_{\substack{a\in\mathcal{A},d\in\mathcal{D}\\ 1\leq m\leq L}}P_a^{(m)}\overline{W}_{a,d}^{(\ell,m)}Q_d^{(\ell)},$$

respectively.

In order to use this form of ARA, Daphne needs an algorithm to solve for the mirroring fixed point solution. Specifically, she must compute the probability distribution P^*. Given P^*, the rest of the solution is simply an optimization problem in which she maximizes with respect to her utilities, rather than the ones that she believe Apollo imputes to her. For that final step, the mirroring fixed point P^* is just an input.

The algorithm used in this problem is based upon the Fictitious Play method of Brown (1949). It incorporates statistical simulation at each iteration step, which is appropriate given the nature of mirroring fixed points: first one optimizes, then takes expectations, then repeats. The main difference between this algorithm and the Fictitious Play scheme is that one only needs to track and update one of the empirical distributions, that for P^*; the other distribution, Daphne's play at each step, is easily found as a (pure-strategy) best response. Therefore, this algorithm estimates only P^* but not $Q^*(\omega)$. Compared to Fictitious Play solutions for Bayes Nash Equilibria, this means that the mirroring fixed point is a computation-friendly solution concept. The iterative algorithm for the ARA mirroring fixed point solution proceeds as follows, where $(\Omega,\mathcal{F},\mathbb{P})$, $a\in\mathcal{A}$ and $d\in\mathcal{D}$ are the primitives:

1. *Initialize.*   Daphne has a probability distribution P_0 over $\mathcal{A}$.
2. *Iterate.*   At step t with distribution P_t, simulate M samples from $(\Omega, \mathcal{F}, \mathbb{P})$.

    2.A   For each sample $\omega \in \Omega$, compute

$$k_t^*(\omega) = \operatorname*{argmax}_{1 \leq k \leq K} \left\{ \left[\widetilde{Y}(\omega) \left(\sum_i p_{ti}\, \boldsymbol{a_i} \right) \right]_k \right\},$$

    where $[\cdot]_k$ denotes the kth element of a vector.

    2.B   Compute the empirical mean:

$$R_t \leftarrow \frac{1}{M} \sum_{\omega \in \Omega} \operatorname*{argmax}_{P \in \mathcal{P}} \ \boldsymbol{e}'_{k_t^*(\omega)} \widetilde{X}(\omega) \left(\sum_i p_i \boldsymbol{a_i} \right),$$

    where $\boldsymbol{e}_k$ denotes the K-dimensional vector with 1 in the kth component and all other components are zero.

    2.C   Update:

$$P_{t+1} \leftarrow \frac{t}{t+1} P_t + \frac{1}{t+1} R_t.$$

    2.D   If $P_t - P_{t+1}$ is sufficiently small with respect to an appropriate metric, stop iterating and set P^* to be P_t. Otherwise, repeat Step 2.

3. *Terminate*   At termination, Daphne chooses the action

$$d^{\mathrm{ARA}} = \operatorname*{argmax}_{d \in \mathcal{D}} \boldsymbol{d}' \boldsymbol{Y} \left(\sum_i p_i^* \boldsymbol{a_i} \right)$$

where $\boldsymbol{Y}$ is her loss matrix.

To illustrate the mirroring fixed-point solution, recall the problem in Fig. 2.3. Suppose that Daphne believes that Apollo thinks this is a zero-sum game, with independent gains for him at each vertex, where each gain is a binomial random variable with $n = 10$ and $p = 0.5$ (e.g., his gain is the number of trucks disabled in an IED attack). For simplicity, we assume that Daphne also believes that Apollo thinks she has the same probability distribution that he does.

From the geometry of the network, it is clear that vertices v_1 and v_2 are equivalent, as are vertices v_3 and v_6 and vertices v_4 and v_5. If Apollo has a single IED, traditional game theory would have him place it at either vertex v_3 or v_6, since the convoy is certain to encounter it. That solution follows from replacing the random payoff at each vertex by its expected value, and then maximizing the minimum gain under Daphne's least favorable routing choice.

But the operations of taking expectations and maximizing the minimum gain do not commute. There is a chance that the random gain at vertex v_3 is small, whereas the gain at v_1 might be large, if Daphne were to select a route that encountered it. For this reason, we expect that the mirror equilibrium solution will put positive probability on vertices v_1 and its equivalent, v_2.

Also, it is clear to Daphne that Apollo does not think she will select a route that encounters vertices v_4 and v_5, since those routes can only increase her risk. If there were any positive probability of an IED at either vertex, then one of the shorter routes would be the dominating strategy. And if there were no probability of an IED at either vertex, then the network diagram shows she would gain no benefit by selecting the longer route.

When the modified version of the Fictitious Play algorithm is applied to this example, where Apollo has a single IED, Daphne finds that she thinks Apollo has probability 0.125 of setting an IED at vertices v_1 and v_2, and probability 0.375 of setting an IED at vertices v_3 and v_6. She believes the chance of an IED at vertices v_4 or v_5 is 0.

The final step in Daphne's analysis is to use this distribution when maximizing her expected utility. Suppose her true loss function is not the binomial Bin$(10, 0.5)$ at each vertex, as she thinks Apollo believes, but rather it is Bin$(10k, 0.3)$, where k is the index of the vertex. In that case she would choose the route $d_2 : S \rightarrow v_1 \rightarrow v_3 \rightarrow v_6 \rightarrow T$, since its expected loss, $3(0.125) + 9(0.375) + 18(0.375) = 10.5$, is minimal.

In this toy example, Daphne was certain that Apollo had only one IED. But that is unreasonable in practice; instead, Daphne will have a subjective distribution over the number of IEDs that Apollo possesses. Suppose she thinks there is a 25% chance that he has just one IED, and a 75% chance that he has two. In that case, she would first use the modified Fictitious Play algorithm for the case with two IED sites, finding that she believes Apollo will select sites (v_3, v_6) with probability 0.4, and sites (v_1, v_3), (v_1, v_6), (v_2, v_3), and (v_2, v_6) with probabilities 0.15. If her true loss function were Binomial$(10k, 0.3)$, then for the case of two IEDs she would again choose route d_2 since its expected loss,

$$(9 + 18)(0.4) + [(3 + 9) + (3 + 18) + (6 + 9) + (6 + 18)](0.15) = 21.6,$$

is the smallest among all the routes.

Both analyses led to the selection of route d_2, and thus Daphne would choose that route, whether there were one or two IEDs. However, her expected loss when she is unsure about the number of IEDs is now $(0.25)(10.5) + (0.75)(21.6) = 18.825$. More generally, if the one-IED case had probability p_1 and she chose route d_1 with expected loss c_1, whereas the two-IED case had probability p_2 and she chose route d_2 with expected loss c_2, then Daphne would select d_1 if and only if $p_1 c_1 \leq p_2 c_2$. That is, she makes the choice that minimizes here expected loss, taking account of her uncertainty about the number of IEDs.

2.3 Comparison of ARA Models

The preceding sections described the key ideas in ARA. The discussion was based upon examples, to provide motivation and build insight into how one might model an adversary's reasoning. This section compares those models, describes their cognitive burdens, and shows how ARA can apply to non-rational behavior through prospect theory.

Consider a discrete simultaneous game in which one player, Daphne, chooses from a finite set of actions $\mathcal{D} = \{d_1, \ldots, d_m\}$ while a second player, Apollo, simultaneously chooses from the finite set of actions $\mathcal{A} = \{a_1, \ldots, a_n\}$. For each pair of choices (d_i, a_j), there is a common random variable ω which determines the utility $u_D(d_i, a_j, \omega)$ that Daphne receives and the utility $u_A(d_i, a_j, \omega)$ that Apollo receives.

To begin, assume that Daphne and Apollo seek to maximize their expected utilities. Given a pair of choices (d_i, a_j), Daphne believes that the density for ω is $p_D(\omega \mid d_i, a_j)$ and Apollo believes it is $p_A(\omega \mid d_i, a_j)$. Then Daphne's expected utility for the pair of choices (d_i, a_j) is

$$\psi_D(d_i, a_j) = \int u_D(d_i, a_j, \omega) p_D(\omega \mid d_i, a_j) d\omega$$

and Apollo's expected utility is $\psi_A(d_i, a_j) = \int u_A(d_i, a_j, \omega) p_A(\omega \mid d_i, a_j) d\omega$. These expected utilities are shown in the following bimatrix.

	a_1	a_2	$\cdots$	a_n
d_1	$(\psi_D(d_1,a_1), \psi_A(d_1,a_1))$	$(\psi_D(d_1,a_2), \psi_A(d_1,a_2))$	$\cdots$	$(\psi_D(d_1,a_n), \psi_A(d_1,a_n))$
d_2	$(\psi_D(d_2,a_1), \psi_A(d_2,a_1))$	$(\psi_D(d_2,a_2), \psi_A(d_2,a_2))$	$\cdots$	$(\psi_D(d_2,a_n), \psi_A(d_2,a_n))$
$\vdots$	$\vdots$	$\vdots$	$\cdots$	$\vdots$
d_m	$(\psi_D(d_m,a_1), \psi_A(d_m,a_1))$	$(\psi_D(d_m,a_2), \psi_A(d_m,a_2))$	$\cdots$	$(\psi_D(d_m,a_n), \psi_A(d_m,a_n))$

If both players know the utility function and probability function of the other, and if they both know that these were common knowledge, then the values in the bimatrix can be used to compute Nash equilibria, typically leading to randomized strategies; see, e.g., Gibbons (1992) or Myerson (1991). However, common knowledge does not hold in the applications considered here and so Nash equilibrium solutions are not applicable.

Without common knowledge, Daphne will need to formulate a probability mass function $p_D(a)$ that represents her beliefs about the probabilities of Apollo's choices. Given that, she selects the action d^* that solves $\mathrm{argmax}_{d \in \mathcal{D}} \psi_D(d)$, where

$$\psi_D(d_i) = \sum_{a \in \mathcal{A}} \psi_D(d_i, a) p_D(a)$$

$$= \sum_{a \in \mathcal{A}} \left[\int u_D(d_i, a, \omega) p_D(\omega \mid d_i, a) d\omega \right] p_D(a). \qquad (2.10)$$

This selection maximizes her expected utility with respect to both her distribution over ω and her distribution over Apollo's choice. Clearly, the strategic difficulty Daphne faces lies in determining $p_D(a)$. The preceding sections described four approaches, each of which is appropriate in some circumstances.

First, suppose Daphne believes her opponent is non-strategic (as discussed in Section 2.2.1). Based on past data and/or expert opinion, she will generate the distribution $p_D(a)$. For example, if Daphne's opponent is Nature and she is choosing among investments in hurricane protection, then she has historical data on hurricane severity and expert opinion on the costs and benefits of different options. Since there is no adversarial strategy, this case is equivalent to a standard risk analysis.

For the special case in which Daphne's opponent chooses actions completely at random, she could use a Dirichlet-multinomial model. Without historical data, she believes that $p_D(a)$ has the Dirichlet distribution with parameter $(\alpha_1, \ldots, \alpha_n)$. But if her opponent has previously chosen action a_j exactly x_j times, then her updated $p_D(a)$ is Dirichlet with parameter $(\alpha_1 + x_1, \ldots, \alpha_n + x_n)$.

In a second special case, assume Daphne thinks that Apollo plays at random, but takes account of the result of the previous play. (For example, in the rock-paper-scissors game, Apollo might favor the choice that won in the preceding round.) Before any rounds are played, Daphne's $p_D(a)$ is the Dirichlet distribution with parameter $(\alpha_1, \ldots, \alpha_n)$. After observing a sequence of plays, Daphne can use the matrix beta prior to learn about the Markov chain of Apollo's choices (Rios and Ríos Insua, 2012). Suppose that, after a (d_i, a_j) choice with discrete outcome η, Apollo has chosen action a_k exactly $x_k^{ij\eta}$ times. Then Daphne's updated distribution, given that the previous play was (d_i, a_j) with outcome η, has the Dirichlet distribution with parameter $(\alpha_1 + x_1^{ij\eta}, \ldots, \alpha_n + x_n^{ij\eta})$. This can be extended to Markov chains with longer memory.

Now consider models for strategic players. When Apollo is strategic, then Daphne (usually) believes that he wants to maximize his expected utility, and seeks the action a^* that solves $\text{argmax}_{a \in \mathcal{A}} \psi_A(a)$, where

$$\psi_A(a_j) = \sum_{d \in \mathcal{D}} \psi_A(d, a_j) p_A(d)$$

$$= \sum_{d \in \mathcal{D}} \left[\int u_A(d, a_j, \omega) p_A(\omega \mid d, a_j) \, d\omega \right] p_A(d). \quad (2.11)$$

So Apollo needs to find $p_A(d)$, his distribution over Daphne's choice.

Daphne does not know the probability functions $p_A(\omega \mid d_i, a_j)$ that Apollo will use, nor does she know his utility function, $u_A(d_i, a_j, \omega)$, nor does she know which $p_A(d)$ he has selected. But, just as she did with the Nash equilibrium seeking adversary, she can model her subjective beliefs about all three quantities through random probabilities and utilities $\{(P_A(\omega \mid d_i, a_j), U_A(d_i, a_j, \omega), P_A(d)\}$. She can now calculate $p_D(a)$ through

$$A \sim \text{argmax}_{a \in \mathcal{A}} \sum_d \left[\int U_A(d, a, \omega) P_A(\omega \mid d, a) \, d\omega \right] P_A(d), \quad (2.12)$$

which enables her to solve (2.10).

In developing the triplet $\{(P_A(\omega \mid d_i, a_j), U_A(d_i, a_j, \omega), P_A(d)\}$, the first two components are usually easier to specify. The $P_A(\omega \mid d_i, a_j)$ does not involve strategy—it is just what Daphne thinks is Apollo's belief about the distribution of the outcome when Daphne selects d_i and Apollo selects a_j. It is random, since Daphne does not know precisely what he thinks, but conventional risk analysis should usually give a reasonable distribution for his beliefs. Similarly, the uncertainty about Apollo's true utility function, $u_A(d_i, a_j, \omega)$, is often small—Daphne has good information about Apollo's objectives and values, so $U_A(d_i, a_j, \omega)$ will have small dispersion. But it is difficult to intuit $P_A(d)$, and the discussion in Section 2.2 considered three methods: Bayes Nash equilibria, level-k thinking, and mirror equilibria.

For example, suppose Daphne thinks Apollo seeks a Nash equilibrium solution (as discussed in Section 2.2.2). Besides $\{(P_A(\omega \mid d_i, a_j), U_A(d_i, a_j, \omega), P_A(d)\}$, Daphne also needs P_D and U_D, the probabilities and utilities that Apollo ascribes to her. These are all defined on a common probability space $(\Theta, \mathcal{F}, \mathcal{P})$ which has atomic elements θ.

Daphne has a distribution for (P_A, U_A), where P_A is the probability distribution that Apollo has for the outcome when the choice pair (d_i, a_j) is selected and U_A is the utility he obtains from that situation. Symmetrically, (P_D, U_D) is her distribution over what Apollo thinks is her probability distribution P_D for the outcome when (d_i, a_j) is selected and U_A is the utility she thinks he believes she obtains from that situation. In most cases, there is little data to guide Daphne in constructing such distributions—she must act as a subjective Bayesian.

For a given $\theta \in \Theta$, the realized random probability distributions and utility functions are (P_A^θ, U_A^θ) and (P_D^θ, U_D^θ). Daphne would calculate the corresponding expected utilities $(\psi_D^\theta(d_i, a_j), \psi_A^\theta(d_i, a_j))$ as:

$$\psi_D^\theta(d_i, a_j) = \int U_D^\theta(d_i, a_j, \omega) P_D^\theta(\omega \mid d_i, a_j) d\omega$$

$$\psi_A^\theta(d_i, a_j) = \int U_A^\theta(d_i, a_j, \omega) P_A^\theta(\omega \mid d_i, a_j) d\omega.$$

For each atom θ, this generates a bimatrix. Daphne would then compute the corresponding Nash equilibria $(d^N(\theta), a^N(\theta))$. These will usually be pairs of distributions over the d_i and a_j choices. When there are multiple equilibria, she should give each equal weight.

Daphne now calculates the expectation with respect to θ, or $p_D^N(a) = E_\mathcal{P}(a^N(\theta))$. Using this expected probability, Daphne would choose the action d^* which maximizes her expected utility; i.e.,

$$d^* = \text{argmax}_{d \in \mathcal{D}} \sum_{a \in \mathcal{A}} \psi_D(d, a) p_D^N(a).$$

Obviously, in most cases there is no closed form solution. Daphne would have to use computational methods to estimate her best decision.

As an alternative to a Nash equilibrium seeking opponent, it may be that Daphne thinks that Apollo is a level-k thinker (as discussed in Section 2.2.3). Daphne needs to find $p_D(a)$ in order to maximize (2.10). In the level-k solution concept, she does this by thinking one level higher than Apollo.

The level-k approach for inferring $p_A(d)$ starts with a non-strategic model for one player, and then finds the best response to the best response to the best response, for k levels. Specifically:

0. If Apollo is a level-0 thinker, then he is non-strategic and Daphne should be a level-1 thinker, acting as described in Section 2.2.1.
1. If Apollo is a level-1 thinker, then he supposes Daphne is a level-0 thinker. Daphne can now infer the analysis he will do, and thereby obtain the $p_D(a)$ she needs. Acting as a level-2 thinker, Daphne plugs her $p_D(a)$ into (2.10) in order to choose her action.
2. If Apollo is a level-2 thinker, he supposes that Daphne is a level-1 thinker, who supposes he is a level-0 thinker. Daphne can now replicate the level-2 analysis just described, but from Apollo's perspective, in order to find $p_D(a)$. Acting as a level-3 thinker, Daphne plugs her $p_D(a)$ into (2.10) in order to choose her action.
3. This ladder of calculation may continue indefinitely. If Apollo is a level-k thinker, Daphne wants to be a level-$(k+1)$ thinker. She replicates the level-k analysis from Apollo's perspective in order to obtain her belief about $p_D(a)$, and then uses that in (2.10) in order to find the action that maximizes her expected utility.

There are two ways for Daphne to terminate this potentially infinite regress. First, she may believe that Apollo reasons to some level k; empirically, people are rarely more level-2 or level-3 thinkers (cf. Lee and Wolpert, 2012; Rothschild, McLay and Guikema, 2012). Second, she may feel that, at some rung in this ladder, she can no longer provide informative subjective beliefs, in which case she terminates the regress with a noninformative distribution (cf. Ríos Insua, Rios and Banks, 2009).

The final model for a strategic opponent finds a *mirror equilibrium* (cf. Section 2.2.4). It is a different way to obtain the $p_D(a)$ that Daphne needs to solve (2.10). She assumes that Apollo is framing the problem in the same way that she does, leading to a pair of coupled equations.

Recall that Daphne is trying to find

$$d^* = \operatorname*{argmax}_{d \in \mathcal{D}} \sum_{a \in \mathcal{A}} \left[\int u_D(d,a,\omega) p_D(\omega \mid d,a) \, d\omega \right] p_D(a)$$

while Apollo seeks

$$a^* = \operatorname*{argmax}_{a \in \mathcal{A}} \sum_{d \in \mathcal{D}} \left[\int u_A(d,a_j,\omega) p_A(\omega \mid d,a_j) \, d\omega \right] p_A(d).$$

Daphne does not know $u_A(d,a_j,\omega)$ nor $p_A(\omega \mid d,a_j)$, and Apollo does not know $u_D(d_i,a,\omega)$ nor $p_D(\omega \mid d_i,a)$.

As a Bayesian, Daphne expresses her uncertainty in terms of probability. Daphne
has distributions for the random quantities $(U_A, P_A(\cdot \mid \cdot), P_A(\cdot))$ which describe her
beliefs about Apollo's utilities and probabilities, as in (2.12), and she has distributions for the random quantities $(U_D, P_D(\cdot \mid \cdot), P_D(\cdot))$ which describe her beliefs about
Apollo's beliefs regarding her own utilities and probabilities.

Suppose that Daphne has a point mass over θ in the probability space $(\Theta, \mathcal{F}, \mathcal{P})$
described previously. So she believes that Apollo will attempt to solve

$$\operatorname*{argmax}_{a \in \mathcal{A}} \sum_{d \in \mathcal{D}} \left[\int U_A^\theta(d, a, \omega) P_A^\theta(\omega \mid a, d) \, d\omega \right] P_A^\theta(d),$$

which provides her guess about his optimal attack a^θ. So, dropping the assumption
of point-mass support, she obtains the distribution over $\mathcal{A}$ given by

$$E_\mathcal{P} \left[\operatorname*{argmax}_{a \in \mathcal{A}} \sum_{d \in \mathcal{D}} \left(\int U_A^\theta(d, a, \omega) P_A^\theta(\omega \mid a, d) \, d\omega \right) P_A^\theta(d) \right],$$

or $E_\mathcal{P}[a^\theta]$, which may be written as

$$\operatorname*{argmax}_{a \in \mathcal{A}} \sum_{d \in \mathcal{D}} \left[\int U_A(d, a, \omega) P_A(\omega \mid a, d) \, d\omega \right] P_A(d).$$

This gives $p_D^M(a)$, Daphne's mirroring solution distribution over Apollo's action.

Symmetrically, given her distribution $p_D^M(a)$ on Apollo's action, Daphne thinks
that Apollo believes that she is trying to solve

$$\operatorname*{argmax}_{d \in \mathcal{D}} \left[\sum_{a \in \mathcal{A}} \int U_D^\theta(d, a, \omega) P_D^\theta(\omega \mid d, a) \, d\omega \right] p_D^M(a).$$

Her mirroring solution is a random action $d^M(\theta)$ which depends upon $(\Theta, \mathcal{F}, \mathcal{P})$.

The probability distributions $p_D^M(a)$ and $\{d^M(\theta), \theta \in \Theta\}$ are consistent if they
jointly satisfy the mirroring condition:

$$d^M(\theta) = \operatorname*{argmax}_{d \in \mathcal{D}} \sum_{d \in \mathcal{D}} \left[\int U_D^\theta(d, a, \omega) P_D^\theta(\omega \mid d, a) \, d\omega \right] p_D^M(a) \tag{2.13}$$

$$p_D^M = E_\mathcal{P} \left[\operatorname*{argmax}_{a \in \mathcal{A}} \sum_{d \in \mathcal{D}} \left(\int U_A^\theta(d, a, \omega) P_A^\theta(\omega \mid d, a) \, d\omega \right) P_A^{\theta M}(d) \right].$$

Finally, given a consistent solution, Daphne should solve for

$$d^M = \operatorname*{argmax}_{d \in \mathcal{D}} \sum_{a \in \mathcal{A}} \psi_D(d, a) p_D^M(a),$$

where p_D^M is obtained as the fixed point in the mirroring analysis (2.13). A mirroring
analysis may be viewed as a way to enforce consistency in Daphne's thinking.

This details the ARA formulation for four different solution concepts: non-strategic choice, the Nash equilibrium, level-k thinking, and the mirror equilibrium. Implementing these different approaches imposes different cognitive loads upon the analyst.

Table 2.3 displays the quantities that Daphne must assess in order to implement a level-k analysis. Row 0 corresponds to the utilities and beliefs of Daphne and Apollo, as perceived by themselves. Subsequently, row k contains the additional utilities and probabilities that Daphne would have to assess in order to perform a level-k analysis. (Recall that, for a level-k analysis, all levels less than k must also be calculated.) The first column contains what Daphne believes are the utility functions that Apollo ascribes to her. The second column contains the probabilities of the outcome, conditional on both her action and Apollo's, that she believes Apollo ascribes to her. The third column contains her opinion of what Apollo thinks is her distribution for he will do. The fourth column contains the utility functions she ascribes to Apollo. The fifth column contains the conditional probabilities of the outcome, given her choice and Apollo's, that she ascribes to Apollo. And, finally, the sixth column contains what she thinks is Apollo's distribution over her choice.

Table 2.3 This table displays the terms used in various kinds of ARA. Each row corresponds to a different level in the "I think that you think that I think ..." style of reasoning.

	1	2	3	4	5	6
0	u_D	$p_D(\cdot \mid d,a)$	$p_D(a)$	u_A	$p_A(\cdot \mid d,a)$	$p_A(d)$
1	U_D^1	$P_D^1(\cdot \mid d,a)$	$P_D^1(a)$	U_A^1	$P_A^1(\cdot \mid d,a)$	$P_A^1(d)$
2	U_D^2	$P_D^2(\cdot \mid d,a)$	$P_D^2(a)$	U_A^2	$P_A^2(\cdot \mid d,a)$	$P_A^2(d)$
$\vdots$	$\vdots$	$\vdots$	$\vdots$	$\vdots$	$\vdots$	$\vdots$

In Table 2.3, different solution concepts use information from different cells:

- Classical game theory requires cells (0,1), (0,2), (0,4), (0,5) and must assume that these are common knowledge.
- Analysis of a non-strategic opponent uses cells (0,1), (0,2), and (0,3); the (0,3) cell is assessed using historical data and/or expert opinion.
- The analysis when the adversary seeks a Nash equilibrium solution uses cells (0,1), (0,2) and (1,1), (1,2), (1,4), and (1,5). These last four cells are needed to infer (0,3).
- The analysis for a level-k opponent requires cells (0,1), (0,2), and:
  - for a level-1 analysis, cells (1,4), (1,5), and (1,6) produce (0,3);
  - for a level-2 analysis, cells (1,1), (1,2), and (1,3) produce (1,6), which, with (1,4), (1,5), then produce (0,3);
  - and so forth for larger k.

- The mirror equilibrium solution uses cells (0,1), (0,2) and imposes a consistency condition between (1,4), (1,5), (1,6), and (1,1), (1,2), and (1,3) to find (0,3).

The main conclusion is that all of these methods entail significant effort. On the other hand, in many real-world situations the decision-maker will have reliable knowledge about columns 1, 2, 4, and 5, and the analysis from there is straightforward.

So far, the discussion has assumed that Daphne and Apollo both seek to maximize expected utility. But there are other criteria which may also be employed within the ARA framework. In the context of terrorism, both psychological research (English, 2009) and logistics requirements (Brown et al., 2006) suggest that terrorists tend to invest resources in order to maximize, or at least acceptably satisfy, some criterion. Although that criterion is not well-specified, it seems improbable that it would be maximum expected utility.

A general method for handling optimization with alternative criteria is *prospect theory* (Wakker, 2010). It includes minimum regret, the Hurwicz criterion, and the Laplace criterion (Luce and Raiffa, 1957), as previously described in Section 2.2.3.

To illustrate how ARA ideas extend to other criteria than maximum expected utility, let $v_A(d_i, a_j, \omega)$ be a value function and let b_A and c_A be weighting functions over Apollo's probabilities $p_A(\omega \mid d_i, a_j)$ and $p_A(d)$. The value function replaces the utility function; it represents the value that Apollo receives when Daphne chooses d_i, he chooses a_j, and the outcome is ω. The weighting functions reflect the empirical fact that people tend to overreact to events with low probability and underreact to events with high probability (Kahneman and Tversky, 1979).

With this machinery, Apollo should seek the solution a^* such that

$$a^* = \text{argmax}_{a \text{ in } \mathcal{A}} \left[\sum_d \int v_A(d,a,\omega)\pi_A(p_A(\omega \mid a,d))\,d\omega \right] \pi_A(p_A(d)),$$

where v_A is a value function and π_A is a weighting function over Apollo's probabilities p_A, as defined in Kahneman and Tversky (1979). Note the formal similarity with equation (2.11).

Daphne does not know these probability distributions, nor the value and weighting functions. She must elicit personal distributions over them, which one models through $(V_A, \Pi_A, P_A(\cdot \mid \cdot), P_A)$. By propagating the uncertainty quantified in these distributions, she induces a probability distribution over Apollo's action space, as in (2.12), through:

$$A^* = \text{argmax}_{a \in \mathcal{A}} \left[\sum_d \int V_A(d,a,\omega)\Pi_A(P_A(\omega \mid a,d))\,d\omega \right] \Pi_A(P_A(d)).$$

This distribution is the $p_D^{PT}(a) = \mathbb{P}(A^* = a)$ required for her to maximize her expected utility in (2.10). Also as before, Daphne could examine a hierarchy of nested decision analyses, in the more complex level-k framework, in order to ultimately elicit $p_D^{PT}(a)$.

Finally, as explained in Section 2.1, all of these methods can be combined in a mixture model in order to express uncertainty about solution concepts and utility functions. When Daphne does not know whether Apollo is non-strategic or a level-1 thinker or is seeking to minimize regret, she may place a subjective probability over all of these possibilities, and then maximize her expected utility with respect to her full uncertainty. These subjective probabilities allow her to combine her solutions through a Bayesian mixture model, as described in Clyde and George (2004) or Hoeting et al. (1999).

Let q_i be the probability that she gives to each of the I opponent models, with $\sum_{i=1} q_i = 1$ and $q_i \geq 0, i = 1, \ldots, I$. Let $p_D^i(a)$ be the probability distribution induced by each opponent model over Apollo's action set, through one of the kinds of ARA described previously. Daphne then combines all these distributions into a single distribution $p_D(a) = \sum_{i=1} q_i p_D^i(a)$, a weighted average for which the weights are her beliefs about the type of reasoning Apollo will use. She should then solve

$$
d^* = \operatorname{argmax}_{d \in \mathcal{D}} \sum_a \psi_D(d,a) \left(\sum_{i=1}^{I} q_i p_D^i(a) \right)
$$

$$
= \operatorname{argmax}_{d \in \mathcal{D}} \sum_{i=1}^{I} q_i \left[\sum_a \psi_D(d,a) p_D^i(a) \right]
$$

to determine her best possible choice. A mixture model increases her computational burden, but it provides a solution that reflects Daphne's honest uncertainty.

Exercises

2.1. Consider the *Battle of the Sexes*, a simultaneous game in which a husband and wife must choose between attending a concert and the ballet without knowing the preference of the other. Both want to go together, but the husband (row-player) prefers the concert and the wife (column-player) prefers the ballet. Their payoff bimatrix is shown below. Find the two pure-strategy Nash equilibria, and describe which equilibrium would be chosen. Also, if the husband and wife both believe the other is non-strategic, what should each do?

	Concert	Ballet
Concert	2, 1	0, 0
Ballet	0, 0	1, 2

2.2. Find the Nash equilibria for the simultaneous game with payoff bimatrix shown below. Compare the equilibrium solutions with those obtained from the maximin and maximax (i.e., maximize the maximum possible gain) solution concepts, and with the solution for the Hurwicz criterion when the weights on the optimistic and pessimistic criteria are both 1/2. Interpret the results in terms of their predictive power.

	Play	Work
Play	2, 2	0, 1
Work	1, 0	1, 1

2.3. Daphne and Apollo must simultaneously choose whether or not to fight each other. Their payoffs are shown below, with Daphne the row player and Apollo the column player. The payoffs associated with (Fight, Fight) are $(x,y) = (1,-2)$ if Daphne is stronger (type $\tau_D = S$) or $(-2,1)$ if she is weaker (type $\tau_D = W$). At the time of the decision, Daphne knows whether she is stronger than Apollo but Apollo does not. Compute a Bayes Nash equilibrium in pure strategies for this game.

		Apollo	
		a_0: Fight	a_1: Peace
Daphne	d_0: Fight	$x,\ y$	$2, -1$
	d_1: Peace	$-1, 2$	$0,\ \ 0$

2.4. The previous problem made an unrealistic assumption of common knowledge about Apollo's belief regarding Daphne's type τ_D. If Apollo believes that Daphne has probability α of being stronger for $\alpha \in (2/3, 3/4)$, would he disclose his true belief to Daphne? If Apollo reports that "$\alpha < 2/3$", does Daphne have any reason to doubt his truthfulness?

2.5. Level-k thinking promotes *social rationality*. Consider the following version of the Prisoner's Dilemma game:

	Cooperate	Defect
Cooperate	10, 10	$-10, 20$
Defect	$20, -10$	0, 0

Compute the Nash equilibrium for this game. Is it a dominated solution? At what value of k does a prisoner who is a level-k thinker realize that the Nash equilibrium is inferior? What prevents players from getting a jointly better solution than the Nash equilibrium? Is there a way to play this game so that a better solution can be reached?

2.6. Recall Example 2.1, concerning the decision made by Captain Hornblower. Rework the analysis for the case in which he is risk-seeking.

2.7. In the smallpox-anthrax problem in Section 2.2.2, show that if $W > Z > X > Y$, then the minimax theorem holds that Daphne should stockpile Cipro with probability $q = (Z-Y)/(W+Z-X-Y)$ and Apollo should undertake a smallpox attack with probability $p = (Z-X)/(W+Z-X-Y)$.

2.8. Consider the *Battle of the Sexes* game described in Exercise 2.1. Suppose the wife knows her husband will use one of the following two solution concepts: (i) he will give equal probability to her possible choices and select the action that maximizes his expected payoff (the Laplace criterion solution) or (ii) he will decide by tossing a fair coin. Build a predictive probabilistic model of the husband's choice. Based on this, what should the wife choose?

2.9. Continuing Exercise 2.8, suppose the wife knows that her husband will seek a Nash equilibrium. This game has three Nash equilibria (including mixed strategies), so the wife does not know which he will play. The wife models the husband's behavior to decide which entertainment she should choose. Does her guess about the Nash equilibrium chosen by the husband reflect aleatory, epistemic, or concept uncertainty?

Chapter 3
Auctions

Chapter 2 described several implementations of adversarial risk analysis (ARA), and compared those with solution concepts used in traditional game theory. This chapter extends that discussion through an in-depth treatment of auctions, a classic problem of practical importance. Specifically, we consider continuous asymmetric, first-price, independent-value, sealed-bid auctions with risk-neutral bidders.

Continuity means that the bids may take any value in an interval, in contrast with the discrete games treated in Chapter 2. Symmetric auctions assume that the values opponents have for the item on offer are randomly drawn from the same (known) distribution, whereas asymmetric auctions allow different opponents to draw from different (known) distributions. In first-price auctions, the highest bidder wins, and pays the amount of that bid. The independent-value condition implies that the private value that one bidder has for an object is not influenced by the private value that other bidders have for that object. The sealed-bid condition ensures that whatever initial information a bidder has about the value distributions of his opponents does not change as the auction proceeds (in contrast with, say, an English auction, where opponents place increasingly higher bids until all but one has dropped out). Finally, risk neutrality implies that each bidder attempts to maximize his expected monetary profit. For concision, we shall use the term "auction" to refer to a continuous asymmetric, first-price, independent-value, sealed-bid auction among risk-neutral bidders.

Auctions of this kind are common, and are especially popular when the commercial value of the item on offer is difficult to determine. They are sometimes used by auction houses, such as Christie's and Sotheby's. In the defense industry, companies make sealed bids on federal contracts, and the *lowest* qualified bidder prevails. But this is a distinction without a difference: the defense contractor's decision problem is formally equivalent to the situation in which the highest bidder wins.

In this chapter, for concreteness, we assume a high-bid auction. Specifically, a lady named Bonnie is bidding against a man named Clyde for a first edition of *The Theory of Games and Economic Behavior*. Also, this auction does not have a reservation price (a secret lower bound set by the owner—if no bid exceeds the reservation price, the book will not be sold).

Auctions easily illustrate aleatory, epistemic, and concept uncertainty. Aleatory uncertainty arises when Bonnie does not have full knowledge of the condition of the book—it could be damaged, which would lower its value, or it might contain marginalia by Lloyd Shapley, which would increase its value. Epistemic uncertainty about the private value of an opposing bidder (i.e., Clyde) can appear in several ways. Perhaps Clyde has better knowledge of the condition of the book; or perhaps the book had been owned by Clyde's thesis advisor, and thus has sentimental value to him. Finally, concept uncertainty occurs when Bonnie does not know what kind of strategic analysis Clyde will perform when calculating his bid.

The following sections consider bidding strategies from several different perspectives, using both classical and ARA techniques. The intent is to highlight the assumptions that are needed and the kinds of solutions that result. When possible, for simplicity, the analysis treats two-person auctions, but the last section discusses three-person auctions, which is sufficient to understand the n-person case.

3.1 Non-Strategic Play

Suppose Bonnie believes that Clyde is non-strategic. In that case, the rule Clyde uses to select his bid for the first edition does not depend upon his analysis of Bonnie's situation. For example, Clyde's rule might be to bid 90% of his true value.

If Bonnie has a distribution F over Clyde's bid, then, under the assumption that her utility function for money is linear, she will maximize her expected utility in a first-price auction by bidding

$$x^* = \text{argmax}_{x \in \mathbb{R}^+} (x_0 - x)F(x), \tag{3.1}$$

where x_0 is Bonnie's true value for the book. To see this, note that her utility (or profit) from a successful bid of x is $(x_0 - x)$, and her personal probability that a bid of x wins the two-person auction is $F(x)$. Thus the right-hand side of (3.1) is just her expected utility when she bids x (cf. Raiffa, Richardson and Metcalfe, 2002). Figure 3.1 illustrates this situation.

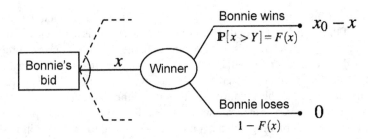

Fig. 3.1 A decision tree in which the possible bids are continuous. The distribution of Clyde's bid is F, so the probability that Bonnie wins with a bid of x is $F(x)$.

As a subjective Bayesian, Bonnie finds the distribution F through introspection, during which she reconciles everything she knows about Clyde with everything she knows about auctions, the market value of first editions of *The Theory of Games and Economic Behavior*, and all other relevant data. Often this is formalized by considering an infinite sequence of hypothetical wagers (De Finetti, 1974). In practice, such precision is impossible, and humans use cognitive shortcuts and approximations when eliciting personal probabilities (cf. O'Hagan et al., 2006).

In this example, a straightforward protocol for Bonnie to obtain her F is to divide the introspection into two parts. First, she puts a subjective distribution G_1 over the value of the first edition to Clyde. Then, she places a subjective distribution G_2 over the fraction of his true value that he bids. The distribution G_1 for Clyde's true value might be approximated by considering the sales prices of other first editions in recent auctions, or the appraisal value by experts, and so forth. Bonnie could adjust G_1 upward if she believes that Clyde puts special value on the book (e.g., she knows that it had been owned by Clyde's thesis advisor). Similarly, to find G_2, Bonnie's distribution for the fraction Clyde bids, she might draw upon knowledge of his success record in previous auctions, or statements he has made in the past, or empirical work in economics on the distribution of underbidding (cf. Case, 2008; Keefer, 1991).

For example, suppose Bonnie describes her epistemic uncertainty about Clyde's true value by a random variable V with distribution G_1 on $(0, \infty)$. She assumes he bids an unknown (and thus, to Bonnie, random) fraction P of that value, for which she has distribution G_2 with support in $[0, 1]$. Then her subjective distribution over $Y = PV$, the amount of Clyde's bid, can be found through a double integral over the shaded region shown in Fig. 3.2. Specifically, when G_1 and G_2 have densities g_1 and g_2, respectively, then

$$F(y) = \mathbb{P}[PV \leq y] = int_0^y \int_0^1 g_1(v)\,g_2(p)\,dp\,dv + \int_y^\infty \int_0^{y/v} g_1(v)\,g_2(p)\,dp\,dv$$

$$= G_1(y) + \int_y^\infty g_1(v)\,G_2(y/v)\,dv. \tag{3.2}$$

This formulation assumes that the true value and the proportional reduction are independent. But if Bonnie thinks that Clyde's non-strategic rule is more complicated (e.g., the proportion P increases as the true value increases), then the analysis is still straightforward, although Bonnie would need to solve a more difficult integral.

It is worth emphasizing that this decomposition of the calculation into a value and a proportion is simply a device for helping Bonnie to develop her personal probability over the fundamental quantity of interest, the bid Y that Clyde makes. She may have other ways to discover F, perhaps through the advice of an informant, or data on Clyde's previous bids.

The following example shows, for the situation in which Clyde bids an unknown fraction of his true value, how the distribution of his bid Y is determined from the distributions G_1 and G_2 that Bonnie needs to assess for her beliefs about the value that the book has to Clyde and the fraction that he will bid, respectively.

Fig. 3.2 Region of integration when the bid is a proportion of the true value.

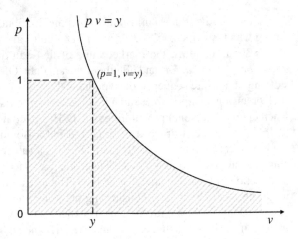

Example 3.1: Suppose Bonnie's personal value for the book on auction is $x_0 = \$150$. She models Clyde's value for the book as a random variable taking values between $0 and $200 with uniform distribution $G_1(v) = v/200$. And she models the distribution for the proportion of his value that he bids as $G_2(p) = p^9$ for $0 \leq p \leq 1$. Then the distribution F of Clyde's bid y is

$$F(y) = G_1(y) + \int_y^{200} g_1(v)G_2(y/v)\,dv = \frac{9}{8}\frac{y}{200} - \frac{1}{8}\frac{y^9}{200^9}$$

for $\$0 \leq y \leq \$200$. Thus Bonnie finds the bid x^* that maximizes her expected utility by solving (3.1). She takes the derivative and sets it to 0, obtaining

$$0 = \frac{d}{dx}\left[(x_0 - x)F(x)\right] = 675 - 9x - \frac{675}{200^8}x^8 + \frac{5}{200^8}x^9.$$

Numerical solution shows her bid should be about half of x_0, or $x^* = \$75$.

In this discussion, concept uncertainty is absent since Bonnie is assumed to believe that Clyde is non-strategic. More precisely, she believes he is the kind of non-strategic player whose bid is proportional to his true value. Her epistemic uncertainty is expressed through the distributions G_1 and G_2, which leads her to F. As posed, there is no aleatory uncertainty, since in this scenario Bonnie knows the value, x_0, that she has for the book.

Now suppose that Bonnie does not know her true value x_0 for the book (e.g., it has not been appraised, or its provenance is uncertain). In that case, its value is a random variable, say X_0, with distribution H. Bonnie wants to make the bid x that maximizes her expected utility $\mathbb{E}_H[(X_0 - x)F(x)] = (\mu - x)F(x)$, where μ is the expected value of X_0. Conveniently, she need not completely specify H; in order to maximize expected utility, all she requires is its mean. (This is a consequence of the risk neutrality assumption, which implies that her utility for money is linear.)

In many situations, aleatory uncertainty is the dominant concern. A defense contractor who bids on a project probably does not know all the costs and difficulties that will arise, and thus does not know exactly what profit would be realized from his bid. This uncertainty can be more important than uncertainty about solution concepts used by opponents, or epistemic uncertainty about the valuations of opponents, especially as the number of opponents increases. In contrast, when each bidder knows the value of the item on offer (e.g., an auction for opera tickets, where each opponent knows his personal utility for opera), then epistemic and concept uncertainty become important. This situation is typical in private value auctions, where each bidder knows his own value, but not those of other bidders. But in common value auctions, for which all bidders have the same (possibly unknown) value for the item on offer, concept uncertainty is likely to dominate the analysis.

3.2 Minimax Perspectives

In a first-price private values auction, the minimax (technically, the maximin) perspective is unhelpful. Bonnie seeks to maximize her minimum utility against the worst possible bid by Clyde. If her preferences are linear in money, and if her true value for the *Theory of Games and Economic Behavior* is x_0, then Bonnie's utility function when she bids x and Clyde bids y is

$$u(x,y) = \begin{cases} x_0 - x & \text{if } x > y \\ 0 & \text{else.} \end{cases}$$

Thus Bonnie solves

$$\max_x \min_y u(x,y)$$

which is 0 for all $x \le x_0$ (since Clyde could bid more than x_0).

If Bonnie knew that Clyde's true value was y_0 with $y_0 < x_0$, then her maximin solution is to bid just a little bit more than y_0. And in the pessimistic limit, as y_0 increases to x_0, Bonnie's bid increases and her profit diminishes to zero. At that limit, any bid x such that $0 \le x \le x_0$ fails to achieve the maximin solution against the worst possible bid by Clyde.

This line of thinking is untenable. Empirical evidence shows that people bid less than their true value (Case, 2008). One of those bidders wins the auction and realizes a positive profit. So a different solution concept is required to analyze bidding.

For example, an ARA solution concept properly leads Bonnie to underbid. She models Clyde's bidding strategy, and then maximizes her expected utility under that model. If she believes that Clyde's bid will have some distribution F, then Bonnie maximizes her expected utility by solving (3.1). When F is continuous with $F(x_0) > 0$, one can show that Bonnie's bid is strictly less than x_0.

Because traditional auctions encourage underbidding, Vickrey (1961) proposed an alternative—the second-price auction, which incentivizes participants to bid their true values. In a second-price auction, the highest bidder wins but pays the amount of the second-highest bid.

To see the underlying logic for the second-price auction, suppose Bonnie bids x and Clyde bids y. Then the gain for Bonnie is $x_0 - y$ if $x > y$, and 0 otherwise. If Bonnie were to bid a value $x^- < x_0$, then one of three outcomes must occur in a second-price auction:

(i) If $y > x_0$, then Bonnie would gain 0 from both a truthful bid x_0 and from her underbid.

(ii) If $y < x^-$, then Bonnie would gain $x_0 - y$ from both a truthful bid x_0 and from her underbid.

(iii) If $x^- \leq y \leq x_0$, then Bonnie would fail to gain $x_0 - y$; she can maximize her minimum gain by increasing her bid to x_0.

Thus Bonnie should not underbid. Similarly, if Bonnie were to bid $x^+ > x_0$, then one of three things must happen:

(i) If $y < x_0$, then Bonnie would gain $x_0 - y$ from both a truthful bid x_0 and from her overbid.

(ii) If $y > x^+$, then Bonnie would gain 0 from both a truthful bid x_0 and from her overbid.

(iii) If $x_0 \leq y \leq x^+$, then Bonnie would lose $y - x_0$; she can minimize her loss by decreasing her bid to x_0.

Thus Bonnie should not overbid. Taken together, these prove that Bonnie ought to bid her true value in a second-price private values auction.

This conclusion for the second-price auction is true whether or not Bonnie adopts an ARA perspective. If Bonnie bids x and Clyde bids y, then Bonnie gains $(x_0 - y)$ if $x > y$ and otherwise gets 0. Bonnie does not know y, so ARA views it as a random variable Y to which she assigns a subjective distribution F. It is convenient to assume that F has density f (but this may be relaxed). Then Bonnie's expected utility from a bid of x^- such that $x^- < x_0$ is

$$\mathbb{E}_F[x^- - Y] = \int_0^{x^-} (x_0 - y)f(y)\,dy$$

and

$$\mathbb{E}_F[x^- - Y] \leq \int_0^{x^-} (x_0 - y)f(y)\,dy + \int_{x^-}^{x_0} (x_0 - y)f(y)\,dy = \mathbb{E}_F[x_0 - Y].$$

The inequality is strict if f is positive on any region in (x^-, x_0). Similarly, if Bonnie bids $x^+ > x_0$, then

$$\mathbb{E}_F[x^+ - Y] = \int_0^{x_0} (x_0 - y)f(y)\,dy + \int_{x_0}^{x^+} (x_0 - y)f(y)\,dy \le \mathbb{E}_F[x_0 - Y],$$

since the term $x_0 - y$ in the second integral is negative. Again, the inequality is strict if $f(y)$ is positive in (x_0, x^+).

However, in some second-price auctions, it is in Bonnie's interest to overbid, but only if there is the prospect of repeated play against the same opponent. Specifically, if Bonnie were confident that Clyde's true value is greater than hers, then overbidding will increase the price he must pay and thus reduce his profit, which will advantage her in future auctions. For example, Boeing knows it will compete with Lockheed Martin for future military contracts. So Boeing wants to ensure that Lockheed Martin makes the smallest possible profit on each contract. If the U.S. Department of Defense awarded contracts through second-price auctions, then Boeing's bidding strategy should treat this game as part of a larger multi-game with repeated play against the same opponent (Camerer, 2003).

It deserves emphasis that in both the first-price and second-price auctions, there are concerns about the applicability of the Nash equilibrium solution concept. Even in the second-price auction there is evidence that real people do not bid as the solution concept directs; instead, they often irrationally underbid (Rothkopf, 2007).

This analysis assumed that Bonnie had no aleatory uncertainty: she knows x_0, her value for the first edition. But, as previously discussed, the logic still holds when the value of the book is a random variable with known mean μ, provided that Bonnie is risk neutral. Since Bonnie seeks to maximize her expected utility, she can replace x_0 by μ in the preceding inequalities, which leads her to bid μ, her expected value for the book.

3.3 Bayes Nash Equilibrium

Much of the research on first-price, sealed-bid, independent-value, rational-bidder auctions has focused on Bayes Nash Equilibrium (BNE) solutions (Krishna, 2010; Klemperer, 2004). There are two cases:

- In the symmetric case, it is assumed that the value each bidder has for the item on auction is a random draw from the same commonly known distribution.
- In the asymmetric case, it is assumed that the value each bidder has for the item on auction is a random draw from a distinct distribution, and that those distributions are known to all bidders.

For the symmetric case, the general solution was found by Vickrey (1961). For the asymmetric case, there remain many open questions; in particular, no current algorithm provably converges to the solution (Fibich and Gavious, 2011).

In the symmetric case, suppose there are n bidders and that the ith bidder has value V_i for the item on offer, where $V_1, \ldots, V_n$ are independent with distribution G. It is assumed that each bidder knows his own value, but not those of his opponents. The strong common knowledge assumption needed for the BNE is that each bidder knows G, and knows that all other bidders also know G. Additionally, in this discussion, assume that all bidders are risk neutral—each seeks to maximize his or her expected profit.

The objective in the symmetric case is to find the bidding function $b(v)$ that maps a value v into a corresponding bid. If one assumes that G is continuous, increasing and differentiable, with support on a compact interval $[L, U]$, then the BNE solution exists and is unique, and $b(v)$ is also continuous, increasing and differentiable (Myerson, 1991, Chap. 3).

Symmetry implies that the equilibrium bidding function $b(\cdot)$ is the same for all players. Suppose that Bonnie values the object at v and, consequently, bids $b(v)$. Then her random profit has expected value

$$[v - b(v)] \mathbb{P}[b(v) \text{ wins}].$$

This is a winning bid if and only if all other players place bids $b(v_i)$ that are less than $b(v)$. Note that G is continuous, so one may ignore the possibility of ties. Since $b(v)$ is strictly increasing, it wins if and only if the values v_i for all other players are less than Bonnie's v. The values are independent, so this happens with probability $G(v)^{n-1}$. Thus her expected profit from bidding $b(v)$ is

$$[v - b(v)] G(v)^{n-1} \tag{3.3}$$

and the same expression holds for all of the other bidders, with their own personal values v_i substituted for Bonnie's value v.

By definition, the equilibrium bid for the ith bidder should be $b(v_i)$. Since this is an optimum of (3.3) and $b(\cdot)$ is continuous, it follows that

$$b(v) = \operatorname{argmax}_w [v - b(w)] G(w)^{n-1}$$

for w in some ball around v. Thus, the derivative of this expression at $w = v$ must be zero:

$$0 = \frac{d}{dw}[v - b(w)] G(w)^{n-1}$$
$$= (v - b(v))(n-1) G'(v) G(v)^{n-2} - b'(v) G(v)^{n-1}.$$

Solving this differential equation shows that for any v in $[L, U]$,

$$b(v) = \frac{\int_0^v z(n-1) G'(z) G(z)^{n-2} \, dz}{G(v)^{n-1}}. \tag{3.4}$$

In general, this expression requires numerical solution.

Example 3.2: Suppose that the value each bidder holds is an independent draw from the distribution $G(v) = v^q$ for $0 \leq v \leq 1$ and $q > 0$. In that case one can find a closed form solution for (3.4):

$$b(v) = v^{-q(n-1)} \int_0^v q(n-1)z^{qn-q}\,dz = \frac{qn-q}{qn-q+1}v.$$

This result shows that as the number n of bidders increases, Bonnie should bid a larger fraction of her true value v, as one would expect. Also, as q increases, Bonnie must bid a larger fraction of her true value.

The ARA perspective has a more natural justification for the BNE solution than the common knowledge assumption. Instead of assuming that everyone knows the common distribution G on the values, it is easier to imagine that Bonnie believes that each of her $n-1$ opponents will draw his true value from G. Bonnie then finds exactly the same solution to the symmetric auction.

The asymmetric auction is more difficult. In general, no expression such as (3.4) exists. When bids are discrete (e.g., they must be whole dollars), then tie-breaking rules are needed to ensure the existence of an equilibrium solution (Lebrun, 1996; Maskin and Riley, 2000a). But when the distributions for the values of the bidders are continuous and differentiable, and one of several possible additional regularity conditions is satisfied, then the bidding functions are unique, continuous, and differentiable (Lebrun, 2006; Maskin and Riley, 2000b). The two most practical regularity conditions are: (1) all value distributions have common support with density that is strictly positive at the lower limit of the support (Lebrun, 1999); or (2) the valuation distributions are locally log-concave at the largest of the lower bounds of the non-common support sets (Lebrun, 2006).

Figure 3.3 shows the MAID that describes the two-person asymmetric auction. The double circle around "Winner" denotes a deterministic node: once both Bonnie's and Clyde's bids are declared, the outcome is non-random.

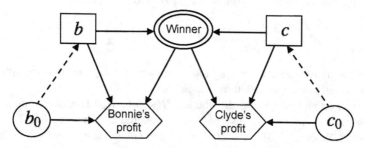

Fig. 3.3 The MAID for a two-person auction. Rectangular decision nodes show the bid that each party makes. Hexagons show the outcome for each bidder given the bids that are placed. Circular nodes indicate that, from the opponent's perspective, the true value is a random variable.

To discuss the asymmetric auction, let F_{IJ} denote what bidder I thinks is the bid that bidder J will place, and let G_{IJ} denote what bidder I thinks is the distribution of bidder J's value. Thus, if Bonnie thinks that Clyde's bid has distribution F_{BC}, her bid should be $b^* = \text{argmax}_{b \in \mathbb{R}^+} (b_0 - b)F_{BC}(b)$ where b_0 is her known true value. And if Clyde thinks that Bonnie's bid has distribution the F_{CB}, then his bid should be $c^* = \text{argmax}_{c \in \mathbb{R}^+} (c_0 - c)F_{CB}(c)$, where c_0 is Clyde's true value. Since neither knows the true value of their opponent, the BNE approach puts commonly known distributions over those values, and solves

$$B^* = \text{argmax}_{b \in \mathbb{R}^+} (B - b)F_{BC}(b) \sim F_{CB} \qquad (3.5)$$
$$C^* = \text{argmax}_{c \in \mathbb{R}^+} (C - c)F_{CB}(c) \sim F_{BC},$$

where $B \sim G_{CB}$ is what Clyde believes is the distribution for Bonnie's true value and $C \sim G_{BC}$ is what Bonnie believes is the distribution for Clyde's true value, and both know what distribution the other has and knows that this is known. If this system of equations has a unique solution, then it determines the F_{BC} that Bonnie needs to calculate her optimal bid, and the F_{CB} that Clyde needs to find his optimal bid. Rarely does (3.5) have a closed-form solution. Kaplan and Zamir (2012) describe some special cases.

Example 3.3: Suppose $B \sim \text{Unif}(0,1)$ and $C \sim \text{Unif}(0,2)$. Then the unique solution to (3.5) is

$$F_{BC}(x) = 4x/(4 - 3x^2) \qquad F_{CB}(y) = 8y/(4 + 3y^2)$$

for $0 \le x, y \le \frac{2}{3}$ (both distributions are 0 for $x, y < 0$ and 1 for $x, y > \frac{2}{3}$).
To verify this, one can find the maximizing B^* and C^* in (3.5) by differentiating and setting the results to 0:

$$B = \frac{F_{BC}(B^*)}{f_{BC}(B^*)} + B^* \qquad C = \frac{F_{CB}(C^*)}{f_{CB}(C^*)} + C^*.$$

For the F_{BC} and F_{CB} that are given, solve for B^* and C^*, obtaining

$$B^* = \frac{4 - 2\sqrt{4 - 3B^2}}{3B} \qquad C^* = \frac{4 - 2\sqrt{4 + 3C^2}}{-3C}.$$

These are both monotone increasing functions, which is logical since the optimum bid should increase with the personal value.
If a random variable W has distribution $H(w)$ and $\theta(\cdot)$ is a monotone increasing transformation with inverse $\theta^{-1}(\cdot)$, then the distribution of $\theta(W)$ is $H(\theta^{-1}(w))$. Since $G_{CB}(b) = b$ for $0 \le b \le 1$ and $G_{BC}(c) = c/2$ for $0 \le c \le 2$, then a little algebra confirms that F_{BC} and F_{CB} solve the system.

When there is no closed-form solution to (3.5) one must use numerical methods. The standard approach is the backshooting algorithm, developed by Marshall et al. (1994). It was later refined by Bajari (2001), Li and Riley (2007), and Gayle and Richard (2008). But Fibich and Gavious (2011) proves that no backshooting algorithm can converge in an epsilon ball around zero. There are additional issues when the value distributions G_{BC} and G_{CB} have one or more crossings (i.e., when one distribution does not stochastically dominate the other, so neither $G_{BC}(x) \geq G_{CB}(x)$ for all x nor $G_{BC}(x) \leq G_{CB}(x)$ for all x). Hubbard, Kirkegaard and Paarsch (2011) proposed a corroborative "visual test" for the accuracy of the numerical solution in the case of multiple crossings, but Au (2014) found errors in the argument. Au proposes a new algorithm, the Backwards Indifference Derivation (BID) scheme, which is successful in cases with known solutions and passes the visual test that backshooting methods sometimes fail. The BID algorithm forms a mesh that discretizes the values of possible bids for each bidder, finds adjacent values between which each bidder is indifferent, and then refines the mesh.

This BNE framework relies upon the common knowledge assumption. For two bidders, the ARA formulation finds the same result through an alternative logic. Instead of common knowledge, Bonnie might reasonably believe that Clyde draws his value from G_{BC} and she also thinks that he believes her value is a draw from G_{CB}. Then Bonnie is led to solve (3.5). But when there are more than two bidders, the ARA perspective opens a larger class of equilibrium problems. Bonnie can model not only what she thinks are the distributions of her opponents' values, but also what she believes are the distributions each opponent has for the values of the other bidders. This topic is further developed in Section 3.6.

3.4 Level-k Thinking

Bayesian level-k thinking is an important family of ARA strategies. The family is diverse, since, at each level, the analyst has many choices regarding how to model the epistemic and aleatory uncertainties. This section applies level-k thinking to auctions (cf. Banks, Petralia and Wang, 2011).

If Bonnie is a level-0 thinker, she bids non-strategically, making no attempt to model her opponents. She might bid 90% of her true value, or place the bid that won a similar book at a recent auction. And the case in which Bonnie is a level-1 thinker was addressed in Section 3.1. She assumed Clyde was non-strategic, and found her best bid given her distribution over his actions.

Things are more interesting when Bonnie is a level-2 thinker. She models Clyde as a level-1 thinker, who believes that Bonnie is a level-0 thinker. Bonnie begins her ARA by developing a subjective distribution F_{CB} to describe what Bonnie believes Clyde thinks is the distribution for her bid. She also needs a subjective distribution G_{BC} for what she believes is Clyde's true value. Finally, she needs to know her own true value b_0 (or, if there is aleatory uncertainty regarding, say, the condition of the book, she needs her expected value, μ).

In this framework, suppose Bonnie believes that Clyde seeks to maximize his expected utility. Clyde knows his true value c_0, and will make the bid c^* such that

$$c^* = \text{argmax}_{c \in \mathbb{R}^+} (c_0 - c) F_B^*(c),$$

where F_B^* is the distribution that Clyde has for Bonnie's bid. Since Bonnie knows neither F_B^* nor c_0, she uses her subjective beliefs to solve the analogous problem:

$$C^* = \text{argmax}_{c \in \mathbb{R}^+} (C_0 - c) F_{CB}(c),$$

where $C_0 \sim G_{BC}$ and F_{CB} is Bonnie's belief about Clyde's belief about the distribution of her bid. Since C_0 is a random variable, then so is C^*; denote its distribution by F_{BC}.

Bonnie has now obtained her belief F_{BC} about the distribution of Clyde's bid, enabling her to solve (3.1). The result is the bid that maximizes her expected utility, where the expectation takes proper account of her uncertainty about both Clyde's true value and his belief about her bid.

Example 3.4: Suppose Bonnie, a level-2 thinker, thinks Clyde believes that her value for the book is a random variable with the uniform distribution on [\$100, \$200]. And further suppose that she thinks Clyde believes that the proportion of her value that she will bid is a random variable with distribution p^9 for $0 \leq p \leq 1$. From Example 3.1 (with the roles reversed), she thinks Clyde's distribution on her bid is $F_{CB}(b) = \frac{9}{8}(b/200) - \frac{1}{8}(b/200)^9$ on [\$0, \$200], and thus his optimal bid is approximately half of his true value.

Bonnie does not know Clyde's true value, but she has a distribution G_{BC} that describes her subjective judgment. Suppose that judgment is that his true value has the triangular distribution on [\$252, \$360] with peak at \$300. Since Clyde should bid 50% of his true value, Bonnie believes that his bid will be a random variable with triangular distribution $F_{BC}(c)$ that is supported on [\$126, \$180] with peak at \$150.

Recall that Bonnie's true value for the book is $b_0 = \$150$. She seeks the bid b^* that maximizes her expected profit, or $(150 - b)F_{BC}(b)$. Simple calculus shows Bonnie should bid \$141.67.

This level-2 solution raises three questions. First, how can one calculate the solution when the assumed distributions are non-trivial? Second, is it reasonable for Bonnie to have such precise opinions about Clyde's beliefs? And third, when should Bonnie proceed to higher levels of thinking?

How Can One Calculate Solutions?
In general there will not be a closed-form solution. The optimum bid must be found numerically, through the following algorithm:

Repeat from $j = 1$ to J:

Sample $c_0^j \sim G_{BC}$

Solve $c_j^* = \text{argmax}_{c \in \mathbb{R}^+} (c_0^j - c) F_{CB}(c)$.

Set $\hat{F}_{BC}(b) = \frac{1}{J} \sum_{j=1}^J I(c_j^* \leq b)$.

Solve $b^* = \text{argmax}_{b \in \mathbb{R}^+} (b_0 - b) \hat{F}_{BC}(b)$.

The first step simulates from what Bonnie thinks is Clyde's value distribution and finds his optimal bid for that random draw. The second step finds the empirical cumulative distribution of his optimal bid. The third solves Bonnie's optimization problem, using the empirical cumulative distribution function of his optimal bid. As J increases, the approximation becomes arbitrarily accurate. If the distributions G_{BC} or F_{CB} are discrete, then the solution may not be unique; adjacent support points (i.e., neighboring values) can have equal expected utility.

Often, F_{CB} is not explicitly available. In that case, the algorithm must be extended. For example, suppose Bonnie believes that Clyde thinks her true value has distribution $G_1(b)$ and that she will bid a random proportion of that value, where the proportion has distribution $G_2(p)$. In that case she can modify the algorithm as follows:

Repeat from $k = 1$ to K:

Sample $v_k \sim G_1$

Sample $p_k \sim G_2$

Set $c_k = p_k v_k$.

Set $\hat{F}_{BC}(c) = \frac{1}{K} \sum_{k=1}^K I(c_k \leq c)$.

Repeat from $j = 1$ to J:

Sample $c_0^j \sim G_{BC}$

Solve $c_j^* = \text{argmax}_{c \in \mathbb{R}^+} (c_0^j - c) \hat{F}_{CB}(c)$.

Set $\hat{F}_{BC}(b) = \frac{1}{J} \sum_{i=1}^J I(y_j^* \leq b)$.

Solve $b^* = \text{argmax}_{b \in \mathbb{R}^+} (b_0 - b) \hat{F}_{BC}(b)$.

As before, the empirical cumulative distribution function $\hat{F}_{BC}$ converges to the distribution F_{BC} as K and J increase. More complicated rules for generating F_{BC} (such as assuming some dependence between V_k and P_k) can be accommodated through modifications of this algorithm.

How Does Bonnie Have Precise Opinions about Clyde's Beliefs?

In real applications, Bonnie would encounter substantial cognitive difficulty in developing her ideas about F_{BC}, and perhaps even G_{BC}. The preceding discussion teased apart that development into two pieces: her belief about Clyde's true value and her belief about the proportion of that value that he bids. But it avoided serious consideration of how to model both distributions. Also, the discussion did not address concept uncertainty. How certain can Bonnie be that Clyde is non-strategic, and that the particular instantiation of his non-strategy is to bid a fraction of his true value?

This difficulty is fundamental—people do not think clearly enough to have fully coherent Bayesian beliefs that incorporate all of their information and intuition. The literature on elicitation of subjective probabilities is extensive and discouraging: experts are overconfident, the framing of the problem matters, mutually contradictory opinions are held, and so forth. O'Hagan et al. (2006), Kahneman (2003), and Garthwaite, Kadane and O'Hagan (2005) are prominent voices in this discussion. Fortunately, in many situations the solution is insensitive to minor errors in the specification of subjective opinion, and sensitivity analysis can flag the cases when greater reflection is required.

There are sensible strategies that can improve Bonnie's assessments. For example, she may not know F_{CB} with confidence, but it is not intellectually overwhelming for her to consider, say, ten fairly distinct choices for it. In examining those ten alternatives for F_{CB}, Bonnie will find that some of them seem more likely to her than others, and she should give those distributions higher probabilities. Then she can combine those by taking the weighted sum of the distributions, with weights corresponding to her probabilities. If done thoughtfully, the resulting distribution will capture much of her judgment. And, of course, the same procedure could be used to formulate her belief about the distribution of Clyde's true value for the book, G_{BC}.

A more sophisticated approach enables Bonnie to express her uncertainty about F_{CB} through a Dirichlet process with central measure F and concentration parameter α. The effect of this would be to increase the variance of F_{BC} above that found from a derivation based on a single F_{CB}. There is a large literature on Bayesian nonparametrics using Dirichlet processes. Müller and Rodriguez (2013) is a short introduction, and Ghosh and Ramamoorthi (2008) provides a more mathematical treatment.

What Level Should One Use?

As previously mentioned, choosing the correct value of k for level-k thinking is an issue. The auction example considered $k = 0, 1, 2$, but one could certainly go higher. Bonnie might attempt to model what Clyde believes Bonnie believes about Clyde, leading to a level-3 analysis (which is straightforward, but the layered reasoning becomes tedious).

Ideally, Bonnie wants to think just one level higher than does Clyde. If she goes beyond that, she is solving the wrong problem, and in general the resulting bidding strategy will be inferior. Thus, selecting the value of k entails uncertainty about precisely which version of the level-k solution concept Clyde is using.

Let p_k denote Bonnie's personal probability that Clyde is a level-k thinker, for $k = 0, 1, \ldots$, such that $\sum p_k = 1$. If $p_k > 0$, Bonnie should do a level-$k+1$ analysis to determine F_{BC}^k, her distribution for Clyde's bid. Then Bonnie combines all of these distributions as a mixture distribution, so that $F_{BC} = \sum p_k F_{BC}^k$. This F_{BC} is the expression of Bonnie's full belief about Clyde's bid, and incorporates her uncertainty about the depth of his strategic thinking. She uses this F_{BC} in (3.1), solving to find her optimal bid.

At some point in the potentially infinite ascent, Bonnie will feel that she no longer has relevant information about Clyde's beliefs. For that value of k, she should assign a uniform distribution over all unknown quantities. A discussion of the use of non-informative distributions to terminate the hierarchy is given in Ríos Insua, Rios and Banks (2009).

3.5 Mirror Equilibria

For auctions, the mirror equilibrium approach is similar to the BNE solution concept. With just two bidders, the key calculation is the same, although the perspective and assumptions are different (cf.0 Banks, Petralia and Wang, 2011). One solves (3.5), but instead of assuming that the distributions are common knowledge, the G_{BC} is Bonnie's belief about the distribution of Clyde's value and the G_{CB} is what she thinks is Clyde's distribution for her value. Then, after deriving F_{BC}, her distribution for Clyde's bid, Bonnie uses her known value b_0 and solves (3.1).

But the mirror equilibrium solution becomes interestingly different from the BNE formulation when the number of bidders is greater than two.

3.6 Three Bidders

Essentially all of the previous discussion addressed games in which there are only two opponents. But an important advantage of ARA is that it enables a more nuanced treatment of many-player games. Specifically, the ARA formulation allows one to frame fresh problems in auction theory when there are more than two bidders, by permitting asymmetric models for how each opponent views the others. We develop the ARA solutions in cases with three opponents for both the level-k thinking solution concept and the mirror equilibrium solution concept. In the following discussion, we now assume that Bonnie is bidding against both Alvin and Clyde to obtain a first edition of the *Theory of Games and Economic Behavior*.

3.6.1 Level-*k* Thinking

If Bonnie is a level-1 thinker, then she assumes that Alvin and Clyde are non-strategic, and there is no novelty in the analysis. She has distributions over the non-strategic bids of each, and chooses her bid according to the maximum of those. Specifically, she has a subjective distribution F_A over Alvin's bid A and a subjective distribution F_C over Clyde's bid C, and she calculates the distribution F of $\max\{A, C\}$. Then she makes the bid

$$b^* = \text{argmax}_{b \in \mathbb{R}^+} (b_0 - b) F(b),$$

where b_0 is her true value for the book.

But now suppose Bonnie is a level-2 thinker. She thinks that Alvin has a belief about the distribution of her bid and also Clyde's bid; similarly, she thinks Clyde has a distribution for her bid and for Alvin's. Recall the previous notation: $F_{IJ}(x)$ is what Bonnie thinks player I thinks is the distribution for player J's bid, and $G_{IJ}(x)$ is her belief about what player I thinks is the distribution for player J's value. Since her level-2 analysis assumes both Alvin and Clyde are level-1 thinkers who believe their opponents are level-0 thinkers, then knowing F_{IJ} directly determines G_{IJ}, as in Example 3.1, where Clyde bids a fraction P of his value V.

The level-2 ARA formulation means that Bonnie thinks Alvin will make the bid $a^* = \max\{a_B^*, a_C^*\}$ for

$$a_B^* = \text{argmax}_{a \in \mathbb{R}^+} (a_0 - a) \mathbb{P}[B^* < a]$$
$$a_C^* = \text{argmax}_{a \in \mathbb{R}^+} (a_0 - a) \mathbb{P}[C^* < a],$$

where a_0 is Alvin's true value, B^* is a random variable whose distribution is Alvin's opinion about Bonnie's bid, and C^* is a random variable whose distribution is Alvin's opinion about Clyde's bid. Bonnie does not know a_0, and she does not know Alvin's distributions for the bids, but as a Bayesian, she has a subjective opinion about these. She regards a_0 as a random variable with distribution G_{BA}, and her best guess is that B^* and C^* have distributions F_{AB} and F_{AC}, respectively.

In order to find F_{AB}, Bonnie uses the fact that Alvin thinks she is a level-0 thinker. He views her as non-strategic, and thus thinks her bid follows some probability distribution, perhaps an unknown proportion of her unknown true value, where both the unknown proportion and the unknown true value can be modeled as random variables. (Of course, she could think that he thinks she follows some other kind of rule, e.g., bidding the last winning bid for similar first editions, or using a random number generator, but she will still always have a subjective distribution over what he believes about the distribution of her bid.)

Thus, Bonnie's opinion about the distribution of Alvin's bid is found by solving

$$A_B^* = \text{argmax}_{a \in \mathbb{R}^+} (A_0 - a) F_{AB}(a)$$
$$A_C^* = \text{argmax}_{a \in \mathbb{R}^+} (A_0 - a) F_{AC}(a)$$

and then assuming that Alvin bids the larger of those two random variables. So his bid is $A^* = \max\{A_B^*, A_C^*\}$.

Similarly, Bonnie belief about Clyde's bid C^* is that it has the distribution of $\max\{C_A^*, C_B^*\}$, where

$$C_A^* = \text{argmax}_{c \in \mathbb{R}^+} (C_0 - c) F_{CA}(c)$$
$$C_B^* = \text{argmax}_{c \in \mathbb{R}^+} (C_0 - c) F_{CB}(c)$$

and C_0 is Clyde's true value, with distribution G_{BC}, since it is unknown to Bonnie. Just as before, Bonnie uses her beliefs about what Clyde thinks about Alvin's non-strategy and her non-strategy to identify F_{CA} and F_{CB}, respectively, and thus finds the distribution of C^*.

Bonnie has calculated her distribution for Alvin's bid A^* and Clyde's bid C^*. Now she should place the bid

$$b^* = \text{argmax}_{b \in \mathbb{R}^+} (b_0 - b) \mathbb{P}[\max\{A^*, C^*\} < b].$$

Generally, this ARA solution will require extensive numerical computation.

One can go further. If Bonnie does a level-3 analysis, she requires two replicates of the level-2 analysis, where Bonnie imagines each opponent is solving his own system of level-2 equations. The nested thinking is complex but straightforward, and the notation must be extended. Let $G_{IJ}(x)$ represent what Bonnie thinks bidder I thinks is the distribution of the value of the book to bidder J, and let G_{IJK} represent what Bonnie thinks bidder I thinks is the distribution that bidder J has for bidder K's value for the book. Similarly, let F_{IJ} represent what Bonnie thinks is the distribution that bidder I has for bidder J's bid, and F_{IJK} represent what Bonnie thinks bidder I thinks is the distribution that bidder J has for bidder K's bid.

Bonnie thinks the level-2 Alvin will reason as follows. First, he thinks Bonnie will make the bid $b^* = \max\{b_A^*, b_C^*\}$ for

$$b_A^* = \text{argmax}_{b \in \mathbb{R}^+} (b_0 - b) \mathbb{P}[A^* < b]$$
$$b_C^* = \text{argmax}_{b \in \mathbb{R}^+} (b_0 - b) \mathbb{P}[C^* < b],$$

where b_0 is Bonnie's true value, A^* is a random variable with distribution F_{BA}, and C^* is a random variable with distribution F_{BC}. Since b_0 is unknown to Alvin, he treats it as a random variable B_0 with distribution G_{AB}. Alvin also does not know know F_{BA} or F_{BC}, but Bonnie believes he thinks A^* has distribution F_{ABA}, and C^* has distribution F_{ABC}. This means that, to Alvin, Bonnie's solutions B_A^* and B_C^* are random variables, and he thinks Bonnie's bid B^* has the distribution of their maximum.

In order to find F_{ABA}, Bonnie thinks level-2 Alvin will model her as a level-1 thinker. That means that he thinks she thinks that his starting point in the level-k

hierarchy is non-strategic. He will have some distribution over what she thinks will be his non-strategic bid, which Alvin must elicit from his personal beliefs. Bonnie does not know what that distribution is, but suppose her subjective belief is that it is H_{ABA}. In that case, Bonnie's best opinion about what Alvin thinks a level-1 Bonnie would bid in order to beat him is

$$B_A^* = \text{argmax}_{b \in \mathbb{R}^+} (B_0 - b) H_{ABA}(b),$$

where B_0 has distribution G_{AB}. Solving this gives F_{ABA}.

Similarly, Bonnie thinks Alvin thinks that her starting point in the level-k reasoning is that Clyde is non-strategic, and thus Alvin must have a distribution over Bonnie's belief about Clyde's bid. Denote Bonnie's best guess about Alvin's distribution for Bonnie's belief about Clyde's bid by H_{ABC}. So Alvin thinks a level-1 Bonnie solves

$$B_C^* = \text{argmax}_{b \in \mathbb{R}^+} (B_0 - b) H_{ABC}(b),$$

where B_0 has distribution G_{AB}, as before. Solving this gives F_{ABC}.

Finally, Alvin should think that Bonnie will bid the maximum of B_A^* and B_C^*. This maximum has distribution F_{AB}.

Similarly, Bonnie thinks Alvin thinks Clyde will bid $C^* = \max\{C_A^*, C_B^*\}$ such that

$$C_A^* = \text{argmax}_{c \in \mathbb{R}^+} (c_0 - c) \mathbb{P}[A^* < c]$$
$$C_B^* = \text{argmax}_{c \in \mathbb{R}^+} (c_0 - c) \mathbb{P}[B^* < c]$$

where c_0 is Clyde's true value, which is unknown to Alvin, and for which Bonnie believes he has distribution G_{AC}. Also, A^* is a random variable that Bonnie thinks has distribution F_{ACA}, and B^* is a random variable that she thinks has distribution F_{ACB}.

Now Bonnie has calculated what she believes Alvin thinks is the distribution of her bid B^* and the distribution of Clyde's bid C^*. So her best guess is that Alvin will make the bid

$$a^* = \text{argmax}_{a \in \mathbb{R}^+} (a_0 - a) \mathbb{P}[\max\{B^*, C^*\} < a].$$

She does not know his value a_0, and thus replaces it with the random variable A_0 with distribution G_{BA}. Solving this new equation provides her distribution F_{BA} for Alvin's bid A^*.

She repeats this reasoning for Clyde instead of Alvin, ultimately obtaining F_{BC}, her distribution for Clyde's bid C^*. Now, Bonnie should make the bid

$$b^* = \text{argmax}_{b \in \mathbb{R}^+} (b_0 - b) \mathbb{P}[\max\{A^*, C^*\} < b].$$

Obviously, implementing the ARA paradigm for the level-k solution concept is intricate—the nested reasoning is difficult for humans to describe, much less perform. But the logic is actually simple, and one can write software that automatically performs these recursions, and thus handles many more than three opponents.

As a final note, when there are more than two bidders, it is possible for different bidders to think at different levels. For example, if Bonnie thinks Alvin is a level-2 thinker but Clyde is only a level-1 thinker, then her analysis might be denoted as level-(3,2) thinking.

3.6.2 Mirror Equilibrium

Now consider the use of the mirror equilibrium solution concept when there are three bidders. This concept assumes that all bidders are solving the problem in the same way, but with possibly different subjective distributions over all unknown quantities.

The two-person system in (3.5) extends so that the basic problem is to solve

$$
\begin{aligned}
A^* &= \text{argmax}_{a \in \mathbb{R}^+} (A_0 - a)F_A^*(a) \\
B^* &= \text{argmax}_{b \in \mathbb{R}^+} (B_0 - b)F_B^*(b) \\
C^* &= \text{argmax}_{c \in \mathbb{R}^+} (C_0 - c)F_C^*(c)
\end{aligned}
\tag{3.6}
$$

from the perspective of each of the players, where $F_I^*(x)$ is what bidder I thinks is the chance that a bid of x will win. Bonnie does not know F_I^*, but she can use ARA to find F_I, which is her belief about what each opponent thinks is the chance that a given bid is successful.

Figure 3.4 may be helpful in following the reasoning. It shows the notation that describes what Bonnie thinks each person believes about the distributions for each of the other bidders' true values. As indicated previously, G_{IJ} is what Bonnie thinks bidder I believes is distribution of the true value for bidder J, and G_{IJK} is the distribution that Bonnie thinks bidder I thinks bidder J has for the true value of the book to bidder K.

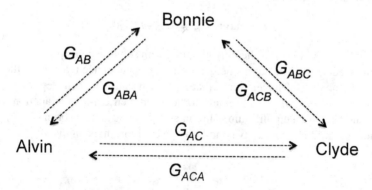

Fig. 3.4 A representation of what Bonnie believes about the opinions held by each of the bidders regarding the value of the book to each the other bidders.

First, she models Alvin's logic. Bonnie thinks he obtains his distribution for her bid by solving (3.6) with $A_0 \sim G_{ABA}$, $B_0 \sim G_{AB}$, and $C_0 \sim G_{ABC}$. Since he, like Bonnie, does not know the true F_I^*, he must develop his own beliefs about them. Here, his F_A is the distribution of the maximum of B^* and C^*, F_B is the distribution of the maximum of A^* and C^*, and F_C is the distribution of the maximum of B^* and C^*. After numerical computation to find the equilibrium solution, he obtains F_{AB}, his belief about the distribution of Bonnie's bid.

Next, Alvin considers Clyde. Bonnie thinks he solves (3.6) with $A_0 \sim G_{ACA}$, $B_0 \sim G_{ACB}$, and $C_0 \sim G_{AC}$. He proceeds as before, and obtains F_{AC}, his belief about the distribution of Clyde's bid. From this, Bonnie thinks his distribution for the probability of winning with a bid of a is F_A, where F_A is the distribution of the maximum of $B \sim F_{AB}$ and $C \sim F_{AC}$.

Bonnie's analysis for Clyde is analogous. To find Clyde's distribution for Bonnie's bid, she thinks he solves (3.6) with $A_0 \sim G_{CBA}$, $B_0 \sim G_{CB}$, and $C_0 \sim G_{CBC}$ to obtain F_{CB}. Similarly, to find Clyde's distribution for Alvin's bid, he uses $A_0 \sim G_{CA}$, $B_0 \sim G_{CAB}$, and $C_0 \sim G_{CAC}$ to obtain F_{CA}. Putting these together, Bonnie thinks that Clyde thinks the probability that a bid of c will win is $F_C(c)$, which is the distribution of the maximum of $A \sim F_{CA}$ and $B \sim F_{CB}$.

Based on this reasoning, Bonnie thinks that Alvin's bid will be

$$A^* = \text{argmax}_{a \in \mathbb{R}^+} (A_0 - a) F_A(a) \sim F_{BA},$$

where $A_0 \sim G_{BA}$. And she thinks Clyde's bid will be

$$C^* = \text{argmax}_{c \in \mathbb{R}^+} (C_0 - c) F_C(c) \sim F_{BC},$$

where $C_0 \sim G_{BC}$. From this, the chance that a bid of b will win is $F_B(b)$, where F_B is the distribution of the maximum of $A^* \sim F_{BA}$ and $C^* \sim F_{BC}$. Now Bonnie uses her known value b_0 and solves

$$b^* = \text{argmax}_{b \in \mathbb{R}^+} (b_0 - b) F_B(b)$$

to obtain her best bid under the mirror equilibrium solution concept.

Some readers may question whether a solution is guaranteed to exist in the mirror equilibrium analysis. The answer is that it must, because at each step, one solves a well-posed Nash equilibrium problem for an asymmetric n-person auction. Lebrun (1999) shows that an equilibrium solution always exists, and Lebrun (2006) proves that, under a mild log concavity condition, the equilibrium is unique.

Exercises

3.1. Suppose n people are bidding to own a Miró in a first-price sealed-bid auction. Each participant has a private valuation for the painting v_i, $i = 1, \ldots, n$: it represents how much the ith bidder is willing to pay. Assume all bidders believe that the others have private valuations that are independent and Unif$[0, 1]$, and that all bidders know each bidder makes this assumption. Consider only bidding strategies of the form $s(v) = \alpha v$ for $\alpha \in [0, 1]$, so that each participant bids a fraction of his valuation. Find a symmetric Bayes Nash equilibrium in this family of strategies.

3.2. Suppose Bonnie is certain that Clyde will bid a fraction P of his true value V for the book *Theory of Games and Economic Behavior*. Also, suppose her distribution over Clyde's value V has a density function with support on $[a, b]$, $0 \le a < b$, and P has distribution supported on $[0, 1]$ and is independent of V. Describe how Bonnie should obtain her distribution for Clyde's bid when $a > 0$. Find Bonnie's optimal bid when her value for the book is $x_0 = \$160$ and she models V with an uniform distribution between $\$100$ and $\$200$ and P with the distribution p^2 for $0 \le p \le 1$.

3.3. In Exercise 3.3, suppose Bonnie models Clyde's value as a triangular distribution supported between $\$100$ and $\$200$ with peak at $\$150$, and his proportional reduction P as a Beta$(20, 10)$. Approximate Bonnie's beliefs about the distribution of Clyde's bid, and obtain her optimal bid when $x_0 = 200$.

3.4. Suppose Bonnie believes there is a positive probability that her opponent's bid will be lower than her value x_0 for the item on offer. Prove that Bonnie's optimal bid x^* against a non-strategic opponent is strictly lower than x_0. Assume that Bonnie's distribution F over her opponent's bid is continuous with $F(x_0) > 0$.

3.5. A Dutch auction (or open-outcry descending-price auction) is an auction in which the seller starts off asking a high price for the item on offer. Then, the price is gradually reduced until a bidder accepts the last announced price. The first bid wins and pays the last price called by the seller. Prove that a Dutch auction has the same optimal bidding strategy as a sealed-bid first-price auction. (Dutch auctions are used when one wants to sell quickly; e.g., bidding on a fishing boat's catch.)

3.6. Bonnie and Clyde are the only bidders for a Juan Gris painting. The auctioneer thinks the item is of high value to both, but he also thinks that each believes the other is an amateur collector who does not value the painting highly. Specifically, the auctioneer thinks Bonnie's and Clyde's valuations are greater than $\$10M$ but each believes the other's valuation is less than $\$1M$. From the auctioneer's perspective, is it smart to have a sealed-bid first-price auction? What auction mechanisms might be better?

3.7. Suppose a $\$100$ bill is offered in a first-price sealed-bid auction between Bonnie and Clyde. Assume bids must be integer multiples of pennies ($\$0.01$). (If both bid the same amount, a coin determines the winner.) Find the Nash equilibrium of this auction. Now suppose Bonnie knows that in real auctions of this kind, participants' bids have the discrete uniform distribution between $\$60$ and $\$100$. How much should she bid?

3.8. In Exercise 3.8, suppose Bonnie is bidding against both Alvin and Clyde. Bids many now be continuous, and she knows that with three bidders, bids are uniformly distributed between $\$70$ and $\$100$. What should Bonnie bid?

Chapter 4
Sequential Games

Sequential games are also amenable to ARA techniques. In sequential games, the participants make decisions over time, usually in alternation. The payoffs could accrue cumulatively during the sequence of play, as with tricks taken in the card game bridge, or the payoff may be determined only at the end of the sequence, as with the checkmate in a game of chess. Often, the payoffs are stochastic. We focus on two-person sequential games with perfect information, meaning that at every stage, each opponent knows the choice that was made by the other.

The key to analyzing sequential games is backwards induction (cf. Aliprantis and Chakrabarti, 2010). Backwards induction can find the Nash equilibria in the sequential Defend-Attack-Defend scenario, an important counterterrorism model (cf. Bier and Azaiez, 2009; Alderson et al., 2011; Rios and Ríos Insua, 2012). This chapter develops the ARA solution to this game, illustrating the ideas in the context of Somali piracy. Finally, there is a discussion of the ARA level-k thinking solution to Emil Borel's classic strategy game *La Relance* (Borel, 1938).

4.1 Sequential Games: The Basics

Sequential games, sometimes called extensive form games, are specified in terms of several elements:

- A set of n opponents (often $n = 2$).
- A game tree, which describes how the game develops, with decision nodes, chance nodes and terminal nodes.
- Decision node labels, indicating which player owns each node.
- A set of moves at each decision node, indicating the choices available to the node owner.
- Distributions at each chance node, describing the values that may be taken by the corresponding random variable.
- Payoffs at the end of each path in the game tree, determining the outcome of the game.

Conceptually, one could treat a sequential game as a simultaneous game, since, in principle, both players might, at the outset, completely prescribe all their choices at each decision node, for all possible choices of the opponent and for all possible values that are realized at the chance nodes. But this is often unrealistic—in the context of chess, the bimatrix for the game would have a row and column for every sequence of lawful moves.

As a simple sequential example, consider a version of the *Entry Deterrence Game*, developed by Dixit (1980). Suppose the Islamic State in Iraq and Syria (ISIS) is deciding whether or not to annex territory in Iraq. If they invade, then Iraq must decide whether to fight or acquiesce. This example assumes that the payoffs are non-random and known to both parties, so the game tree is actually a decision tree, since there are no chance nodes. Figure 4.1 shows this decision tree, with payoff pairs at the termini for each decision path. The hypothetical payoffs are in billions of dollars; the first payoff is what ISIS receives, and the second is what Iraq receives. The intuition is that if ISIS invades and Iraq chooses to fight, then ISIS will eventually win and keep the territory, which is worth $5 billion, but both parties pay a cost for conflict.

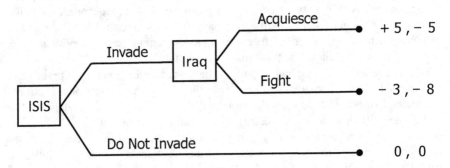

Fig. 4.1 An illustration of a game for which one of the equilibria, (Not Invade, Fight), is not a subgame perfect Nash equilibrium.

This game is sufficiently simple that it can be written in simultaneous form, as shown in Table 4.1. The rows are the choices ISIS can make, and the columns are the choices that Iraq can make. One can see that there are two Nash equilibria: one of them is the impossible (Not Invade, Fight) and the other is (Invade, Acquiesce). In both cases, neither opponent gains from a unilateral change in its choice.

Table 4.1 The payoff table for an ISIS-Iraq game. The rows indicate whether the first player, ISIS, decides to invade or not. The columns indicate whether Iraq decides to fight or not.

	Fight	Acquiesce
Invade	(−3, −8)	(5, −5)
Don't Invade	(0, 0)	(0, 0)

In sequential games, not all Nash equilibria are reasonable solutions. These unreasonable solutions motivate the subgame perfect Nash equilibrium concept, which precludes the problem. These solutions require that each player's strategy be optimal, not only at the start of the sequential game, but also at every decision point.

In the ISIS-Iraq game, there are two Nash equilibrium solutions. While Iraq would prefer the (Not Invade, Fight) solution, and thus will threaten to fight ISIS in order to dissuade an invasion, in practice the decision is made sequentially rather than simultaneously. If ISIS were to invade, the only rational move for Iraq is to cede the territory. ISIS knows this, and will surely invade. Therefore the (Invade, Acquiesce) solution is the only reasonable outcome.

Here are three relevant definitions:

1. A player's strategy is *sequentially rational* if it maximizes his expected utility, conditional on the information he has at the time he chooses an action.
2. In a game tree, a node creates a *subgame* if neither the node nor any of its successors contains information pertinent to nodes that are not successors of the node. The tree defined by the node and its successors is a tree for a subgame.
3. A solution to a sequential game is a subgame perfect Nash equilibrium if it finds a Nash equilibrium in each of its subgames.

For the ISIS-Iraq game, consider the subgame formed by its decision to fight or acquiesce. That subtree has two choices, and clearly the $(5, -5)$ solution is better for Iraq than the $(-3, -8)$ solution; i.e., it is is the sequentially rational decision. So one can replace the Iraqi decision node by the payoff $(5, -5)$. Now ISIS has two choices, $(0, 0)$ or $(5, -5)$, and clearly should choose the latter.

Subgame perfect equilibria are computed by backwards induction, which is the analog of rolling-back trees in decision analysis (cf. Osborne, 2004) or using dynamic programming in settings with multiple players. Harrington (2014) outlines backwards induction for sequential games with perfect information as follows:

1. For each of the final decision nodes, find the optimal choice.
2. At each of those decision nodes, replace the part of the tree beginning with that decision node with the associated expected payoffs, assuming optimal play.
3. Repeat steps 1 and 2 for this reduced game until the initial decision node is reached.

Implicit in this algorithm is the idea that a player anticipates that other players will act optimally. One can show, as in Aliprantis and Chakrabarti (2010), that any outcome selected by backwards induction is a subgame perfect Nash equilibrium.

As an important example of sequential games, consider the Defend-Attack-Defend game (Brown, Carlyle and Wood, 2008; Parnell, Smith and Moxley, 2010). In this game the Defender (she) first deploys her defensive resources. The Attacker (he) observes the deployment and chooses his attack. Then, the Defender attempts to recover from the attack as best she can.

Figure 4.2 shows coupled influence diagrams in the MAID, with a shared uncertainty node Ⓢ , and a game tree, both representing the Defend-Attack-Defend
model. Nodes $\boxed{D_1}$ and $\boxed{D_2}$ correspond to the Defender's first and second decisions,
d_1 and d_2, respectively, and node $\boxed{A}$ represents the Attacker's decision, a. The possible choices are $d_1 \in \mathcal{D}_1$, $a \in \mathcal{A}$ and $d_2 \in \mathcal{D}_2$.

As written, this model assumes that the only relevant uncertainty is the success
level S of the attack, which depends probabilistically on $(d_1, a) \in \mathcal{D}_1 \times \mathcal{A}$. The payoff
to the Defender depends on (d_1, s, d_2): the cost of her initial defense, the success of
the attack, and the success of the recovery effort. The payoff to the Attacker depends
on (a, s, d_2): the effort in mounting his attack, its success, and the success of the
recovery effort. The model may be easily generalized so that the outcome of the
recovery effort is also a random variable.

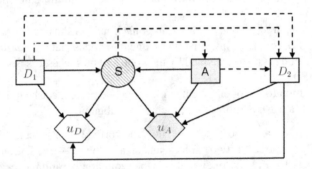

(a) Multi-agent influence diagram for the Defend-Attack-Defend game.

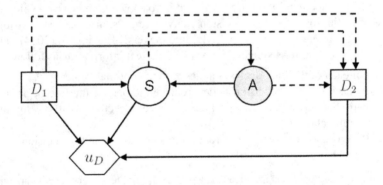

(b) The Defend-Attack-Defend game tree.

Fig. 4.2 Two views of the Defend-Attack-Defend model. The MAID shows the information available to each opponent at each stage of the game, and the relationships between utilities, decisions,
and chance. The game tree shows the sequence of play, which is helpful in backwards induction.

Game theory needs the Defender to know the Attacker's utilities and probabilities, and the Attacker to know the Defender's utilities and probabilities, and for both to know that these are common knowledge. Often, the utility functions and probability assessments depend upon all of d_1, s, a, and d_2 (i.e., the utility and probability are affected not only by the chance outcome, but also by the choices of each opponent). However, the following discussion simplifies things a bit and assumes the utilities and probabilities depend only upon the outcome of the opponent's decision, but not also on the decision itself (as is reasonable for both examples in this chapter). In such cases, denote the Defender and Attacker utility functions by $u_D(d_1, s, d_2)$ and $u_A(a, s, d_2)$, respectively, and their probability assessments about the success of attack by $p_D(S = s \mid d_1, a)$ and $p_A(S = s \mid d_1, a)$, respectively.

Using backwards induction, at node D_2 for the game tree in Figure 4.2(b), the Defender's best response after observing $(d_1, s) \in \mathcal{D}_1 \times S$ is

$$d_2^*(d_1, s) = \text{argmax}_{d_2 \in \mathcal{D}_2} u_D(d_1, s, d_2). \tag{4.1}$$

Under the common knowledge assumption, the Defender's behavior at $\boxed{D_2}$ will be anticipated by the Attacker. Thus, at node $\textcircled{S}$, the Defender's expected utility associated with each $(d_1, a) \in \mathcal{D}_1 \times A$ is

$$\psi_D(d_1, a) = \int u_D(d_1, s, d_2^*(d_1, s)) \, p_D(S = s \mid d_1, a) \, ds, \tag{4.2}$$

and the Attacker's expected utility is

$$\psi_A(d_1, a) = \int u_A(a, s, d_2^*(d_1, s)) \, p_A(S = s \mid d_1, a) \, ds,$$

and these are known to both opponents. The Attacker now finds his optimal attack at node A, after observing the Defender's move $d_1 \in \mathcal{D}_1$, by solving

$$a^*(d_1) = \text{argmax}_{a \in A} \psi_A(d_1, a).$$

Knowing this, the Defender can find her maximum expected utility decision at node $\boxed{D_1}$ through

$$d_1^* = \text{argmax}_{d_1 \in \mathcal{D}_1} \psi_D(d_1, a^*(d_1)),$$

which gives the full solution to the game.

Assuming common knowledge, rational players, and perfect information, game theory prescribes that the Defender should choose $d_1^* \in \mathcal{D}_1$ at node $\boxed{D_1}$, the Attacker should choose attack $a^*(d_1^*) \in A$ at node $\boxed{A}$ after observing d_1^*, and, finally, Defender, after observing $s \in S$, should choose $d_2^*(d_1^*, s) \in \mathcal{D}_2$ at node $\boxed{D_2}$.

4.2 ARA for Sequential Games

ARA can be used in sequential games, and it allows one to drop the assumption that all utilities and probabilities are common knowledge. To describe this alternative to the previous solution, consider again the Defend-Attack-Defend model.

From the ARA perspective, the Attacker's decision at node A is uncertain to the Defender, and she must model her uncertainty through a random variable. This is reflected in the influence diagram and the game tree in Figure 4.3, where the Attacker's decision node $\boxed{A}$ has been converted to the chance node $\bigcirc\!\!\!A$. The Defender needs to assess $p_D(A \mid d_1)$, her predictive distribution about which attack will be chosen at node A for each $d_1 \in \mathcal{D}_1$. Additionally, she must make the (more standard) assessments about $u_D(d_1, s, d_2)$ and $p_D(S \mid d_1, a)$.

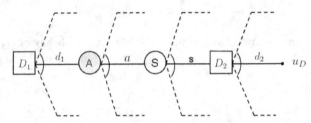

(a) The influence diagram for the ARA solution of the Defend-Attack-Defend game.

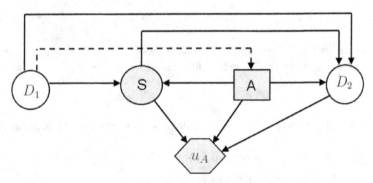

(b) The game tree for the ARA solution of the Defend-Attack-Defend game.

Fig. 4.3 The Defender's decision problem, represented as an influence diagram and as a game tree.

Given these, the Defender works backward in the tree in Figure 4.3(b) to solve her problem. At node $\boxed{D_2}$ she computes her maximum utility action $d_2^*(d_1, s)$ for each $(d_1, s) \in \mathcal{D}_1 \times S$, as in (4.1). Then, at node $\bigcirc\!\!\!S$, she finds her expected utility $\psi_D(d_1, a)$ for each $(d_1, a) \in \mathcal{D}_1 \times A$, as in (4.2). The next step employs her subjective probability assessment of what the Attacker will do, $p_D(A \mid d_1)$, in order to compute her expected utility at node Ⓐ for each $d_1 \in \mathcal{D}_1$:

$$\psi_D(d_1) = \int \psi_D(d_1, a) \, p_D(A = a \mid d_1) \, da.$$

Finally, the Defender finds her maximum expected utility decision at $\boxed{D_1}$ as

$$d_1^* = \text{argmax}_{d_1 \in \mathcal{D}_1} \psi_D(d_1).$$

Thus, backwards induction shows that the Defender's best strategy is to choose first d_1^* at node $\boxed{D_1}$, and later, after observing $s \in S$, choose $d_2^*(d_1^*, s)$ at node $\boxed{D_2}$.

As usual, ARA requires the decision maker to assess $p_D(A \mid d_1)$. This could be done through some form of risk analysis (cf. Ezell et al., 2010), or by developing a model for the strategic analysis of the opponent. Given $p_D(A \mid d_1)$, the analysis of the Attacker's decision problem, as seen by the Defender, is shown in Figure 4.4, where the Attacker's probabilities and utilities are assessed by the Defender. That assessment may be based upon historical data and expert opinion. If salient information is not available, then the Defender may choose to use a noninformative distribution for $p_D(A \mid d_1)$.

In order to elicit $p_D(A \mid d_1)$, the Defender must assess $u_A(a, s, d_2)$, $p_A(S \mid d_1, a)$, and $p_A(D_2 \mid d_1, a, s)$. In general, she does not know these quantities, and represents her uncertainty through a joint distribution F on $U_A(a, s, d_2)$, $P_A(S \mid d_1, a)$, and $P_A(D_2 \mid d_1, a, s)$. These distributions might be elicited in various ways, such as the ordinal judgment procedure proposed by Wang and Bier (2013). She then solves her perception of the Attacker's decision problem using backward induction over the game tree in Figure 4.4(b), propagating her uncertainty, encoded by F, to get distributions over the random action $A^*(d_1)$ for each d_1. Specifically, if all choice sets are continuous, she solves as follows:

- At decision node $\boxed{D_2}$, compute the random expected utilities

$$(d_1, a, s) \rightarrow \Psi_A(d_1, a, s) = \int U_A(a, s, d_2) \, P_A(D_2 = d_2 \mid d_1, a, s) \, dd_2.$$

- At chance node Ⓢ, compute the random expected utilities

$$(d_1, a) \rightarrow \Psi_A(d_1, a) = \int \Psi_A(d_1, a, s) \, P_A(S = s \mid d_1, a) \, ds.$$

- At decision node $\boxed{A}$, compute the random optimal initial decision

$$d_1 \rightarrow A^*(d_1) = \text{argmax}_{a \in \mathcal{A}} \Psi_A(d_1, a).$$

Thus, the Defender's predictive distribution over attacks, conditional on her first defensive decision d_1, is given by

$$\int_0^a p_D(A = x \mid d_1) \, dx = \Pr[A^*(d_1) \leq a].$$

(If the choice sets are not continuous, the Defender would reason similarly, but replace integrals with sums and find the predictive distribution over a set.)

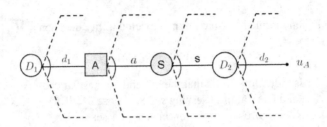

(a) Influence diagram for the Defender's view of the Attacker's problem.

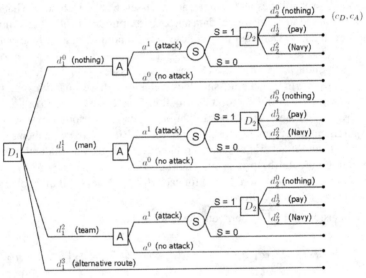

(b) Game tree for the Defender's view of the Attacker's problem.

Fig. 4.4 The Defender's view of the Attacker's decision problem.

When the beliefs are complex, the Defender's predictive distribution can be approximated using Monte Carlo. Specifically, for each $d_1 \in \mathcal{D}_1$, the following algorithm finds the Defender's belief about the probability of each attack.

Do, for $i = 1, ..., N$:

Draw $\left(U_A^i(a, s, d_2), P_A^i(S \mid d_1, a), P_A^i(D_2 \mid d_1, a, s)\right) \sim F$.

At chance node D_2, compute

$$(d_1, a, s) \rightarrow \psi_A^i(d_1, a, s) = \int U_A^i(a, s, d_2) \, P_A^i(D_2 = d_2 \mid d_1, a, s) \, dd_2.$$

At chance node S, compute

$$(d_1, a) \rightarrow \psi_A^i(d_1, a) = \int \psi_A^i(d_1, a, s) \, P_A^i(S = s \mid d_1, a) \, ds.$$

At decision node A, compute

$$d_1 \rightarrow a_i^*(d_1) = \operatorname{argmax}_{a \in A} \psi_A^i(d_1, a).$$

Approximate $\int_0^a p_D(A = x \mid d_1) dx$ by $\#\{a_i^*(d_1) : a_i^*(d_1) \leq a\}/N$ where $\#\{\cdot\}$ is the cardinality of the set.

Given F, the Defender's assessment of $p_D(A \mid d_1)$ is straightforward. In many situations, the random utility $U_A(a, s, d_2)$ and the random probability $P_A(S \mid d_1, a)$ in F are relatively simple to elicit. However, the assessment of $P_A(D_2 \mid d_1, a, s)$ within F can be problematic; the Defender should exploit any information she has on how the Attacker analyzes her decision problem. She may think he seeks a Nash equilibrium, or uses level-k thinking; if she has little insight, she can represent that through some high-variance or non-informative distribution. The situation is similar to that discussed in Section 2.3.

4.3 Case Study: Somali Pirates

As a concrete example of sequential ARA, consider the decisions made in defending a ship from piracy. Between 2005 and 2011, pirates threatened shipping in the Gulf of Aden. No ship within several hundred miles of the Somali coast was safe, and, until recent successful countermeasures, it was a major issue in international security.

Economic and political changes have led many Somali fishermen to take up piracy, and infrastructure evolved to support this new enterprise (cf. Carney, 2009). The infrastructure had multiple agents: village elders who act as the de facto local government, Somali businessmen who invest in piracy, and negotiators who broker ransoms for captured ships and crews. So an analyst can model that the pirates as pragmatic businessmen who pursue their goals strategically.

A typical attack is undertaken by a small groups of about ten men in fast boats which depart from a mother-ship. If successful, about 50 pirates are left on-board to pilot the captured ship into harbor, while another 50 or so pirates provide logistical support from the shore. The goal is ransom rather than theft; it is more profitable, and the pirates reinvest part of their gains in equipment and training.

First consider anti-piracy strategy from the perspective of a ship owner, structuring the problem as a game tree. This leads to a sequential Defend-Attack-Defend game between the Owner and the Pirates, following the formulation in Sevillano, Ríos Insua and Rios (2012).

For the first defensive decision, the Owner may choose among many options, including various levels of on-board armed security and selecting an alternate (longer) route that avoids the Gulf of Aden. Next, the Pirates, who have a network of spies that provide information about security, cargo and crew, respond to the Owner's initial decision by either attacking or not attacking the ship. If the pirate attack is successful, the Owner will have to decide how much to pay in ransom, or perhaps she will hire armed forces to re-seize the ship.

Specifically, assume that the Owner can select one of the following four defensive actions (i.e., elements of $\mathcal{D}_1$):

d_1^0:   Do nothing, i.e., no defensive action is taken.
d_1^1:   Hire an armed guard to travel on the ship.
d_1^2:   Hire a team of two armed guards for the ship.
d_1^3:   Use the Cape of Good Hope route, not the Suez Canal.

Once the Owner has made her initial choice, the Pirates observe it and decide whether or not to attack (a^1 and a^0, respectively, in $\mathcal{A}$). An attack either results in the ship being hijacked ($S = 1$) or not ($S = 0$), with probabilities that depend on the Owner's initial choice. If the ship is hijacked, then the Owner has three possible responses (elements of $\mathcal{D}_2$):

d_2^0:   Do nothing, i.e., refuse to pay the Pirates' ransom.
d_2^1:   Pay a ransom, thus recovering the ship and crew.
d_2^2:   Pay the Navy (specifically, Combined Task Force 150) to recapture the
          ship and crew.

Obviously, this framing of the decision problem is only an approximation to the true complexity of the application, but it captures the salient elements.

The asymmetric game tree shown in Figure 4.5 represents the sequence of decisions and outcomes faced by the Owner and the Pirates. The nodes $\boxed{D_1}$ and $\boxed{D_2}$ correspond to the Owner's first and second decisions, respectively, the node $\boxed{A}$ represents the Pirates' decision, and the chance node S represents the outcome of the attack (if undertaken). The pair (c_D, c_A) represents the consequences to the Owner and the Pirates, respectively, from the corresponding sequence of decisions and outcomes.

The Owner's sequential ARA decision problem is mapped as the game tree in Figure 4.6, in which the Pirates' decision node $\boxed{A}$ has been replaced by the chance node Ⓐ. This shows that the Pirates' decision is unknown in advance to the Owner,

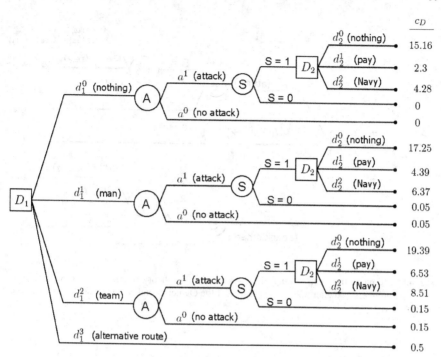

Fig. 4.5 Game tree for the Somali pirates case study.

forcing the bulk of her modeling to focus on assessments of probabilities over the Pirates' actions conditional on her own. Thus, to solve her decision problem, she needs to assess $p_D(A \mid d_1)$, her predictive probability of an attack given each $d_1 \in \mathcal{D}_1$. Additionally, she needs to make assessments $p_D(S \mid d_1, a^1)$ and $u_D(c_D)$, with c_D representing her monetary cost equivalent from the multi-attribute consequences associated with each branch in the tree. These assessments are easier to make because they are non-strategic, and also because in this application there is relevant historical information.

For the Owner, the possible consequences are loss and ransom of the ship, the costs associated with defense preparation, the costs associated with a possible battle, and the profit from a voyage that is either not attacked or which successfully repels armed boarders. These costs are not all commensurate—an attack, successful or unsuccessful, could include loss of life on either side. This analysis assumes that the Owner puts no value on the life of a Somali pirate, and uses the value of a statistical life for each member of the crew (including any guards that may be hired). Viscusi and Aldy (2003) reviews various methods for estimating the value of a statistical life (e.g., the total income over the remainder of an expected lifespan); this analysis uses the one in Martinez and Mendez (2009).

More specifically, the Owner's direct costs for the defensive actions in $\mathcal{D}_1$ are as follows:

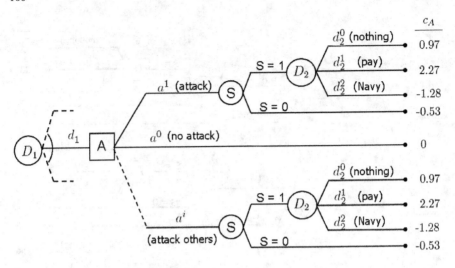

Fig. 4.6 The game tree representing the Owner's decision theory problem for the Somali pirates case. The numbers in the terminal nodes represent the Owner's total cost for that path, as computed in Table 4.1.

- €0, if the Owner hires no guards (i.e., d_1^0).
- €0.05M, if the Owner hires one guard (i.e., d_1^1). This includes the six-month salary, equipment, travel expenses, and so forth.
- €0.15M, if the Owner hires two guards (i.e., d_1^2). This includes the six-month salaries for two armed guards, with better equipment and associated expenses.
- €0.5M, if the Owner chooses to circumnavigate the Cape of Good Hope (i.e., d_1^3). The increased distance entails higher operating expenses and opportunity costs.

Similarly, the Owner's costs for the defensive actions in $\mathcal{D}_2$ are:

- €0, if the ship is attacked and the Owner pays no ransom (i.e., d_2^0). (This is the direct cost; the loss of the ship and cargo is accounted for separately in Table 4.1.)
- €2.3M, if the ship is successfully attacked and the Owner pays a ransom (i.e., d_2^1). The €2.3M figure is the average of recent ransoms reported by Carney (2009): €2.2M for *Le Ponant*; €2M for *MT Stolt Melati 5*; €1.1M for *MT Stolt Valor*; €3M for *Sirius Star*; and €3.2M for *MV Faina*. The ransom payment could be treated as a random variable, but for simplicity it is assumed to be fixed.
- €0.2M, if the ship is attacked and the Owner calls for the Navy (i.e., d_2^2). This figure is based on fees for military intervention using the international coalition ships already deployed in the area.

Besides these costs, some paths in the tree entail deaths. The Owner's direct cost for these depends upon insurance and litigation but the ARA should take deaths into account.

For this example, the analysis assumes that if an attack is successfully repelled ($S = 0$), then no lives are lost. But in a successful attack ($S = 1$), the analysis assumes that all armed guards are killed and, depending on the chosen response at the node $\boxed{D_2}$, there may be additional fatalities: (1) if the Owner does not ransom the ship, the angry Pirates kill four of the crew; (2) if the Owner pays the ransom, no one else dies; (3) if the hijacked ship is rescued by the Navy, there are two more deaths. A slightly more complex analysis would properly treat the number of deaths and their costs as random variables, but this discussion omits that and follows Martinez and Mendez (2009) in fixing the statistical value of a (Spanish) life at €2.04M. In similar simplification, we assume the depreciated value of the ship and its cargo is €7M.

Table 4.2 summarizes the costs c_D for the Owner that are associated with each scenario (i.e., each path in the tree shown in Figure 4.5). Clearly, if there is no attack, then $S = 0$.

Table 4.2 The Owner's costs associated with different tree paths.

D_1	S	D_2	Ship loss	Action costs	Lives lost	c_D
do nothing	$S = 1$	don't pay	1	0 + 0	0 + 4	15.16
do nothing	$S = 1$	pay ransom	0	0 + 2.3M	0 + 0	2.30
do nothing	$S = 1$	call Navy	0	0 + 0.2M	0 + 2	4.28
do nothing	$S = 0$		0	0	0	0.00
hire guard	$S = 1$	don't pay	1	0.05M + 0	1 + 4	17.25
hire guard	$S = 1$	pay ransom	0	0.05M + 2.3M	1 + 0	4.39
hire guare	$S = 1$	call Navy	0	0.05M + 0.2M	1 + 2	6.37
hire guard	$S = 0$		0	0.05M	0	0.05
hire team	$S = 1$	don't pay	1	0.15M + 0	2 + 4	19.39
hire team	$S = 1$	pay ransom	0	0.15M + 2.3M	2 + 0	6.53
hire team	$S = 1$	call Navy	0	0.15M + 0.2M	2 + 2	8.51
hire team	$S = 0$		0	0.15M	0	0.15
d_1^3 (alternative route)			0	0.5M	0	0.50

If the Owner has constant absolute risk aversion, then her utility function has the form $u_D(c_D) = 1 - \exp(-\alpha \times c_D)$, with $\alpha > 0$ (see Section 2.2.1). Constant risk aversion implies that when choosing between a guaranteed payment and a gamble, her choice is the same for any constant multiple of both the payment and the expected value of the gamble. As a sensitivity analysis, this ARA considers her optimal decision for $\alpha \in \{0.1, 0.4, 1, 2, 5\}$.

Based on historical information (Carney, 2009), the ARA assumes that the Owner believes that an attack will be successful when no armed guards are hired is 0.4, or

$p_D(S = 1 \mid a^1, d_1^0) = 0.40$. It also assumes that with one guard, the chance that the attack will be successful is 0.1 (i.e., $p_D(S = 1 \mid a^1, d_1^1) = 0.1$), and with two guards, the chance is 0.05 (i.e., $p_D(S = 1 \mid a^1, d_1^2) = 0.05$). Later, we vary these probabilities to determine their influence on the analysis.

In order to implement an ARA, the Owner must estimate the probability of attack conditional on each of her initial defensive choices. One could try to estimate these from historical data—Carney (2009) puts the overall probability of attack at about 0.005. But this does not account for the defensive choices, nor the intelligence network used by the Pirates to identify ships with valuable cargo and vulnerabilities, so using 0.005 as the attack probability implies a non-strategic opponent. Instead, the Owner performs a level-2 analysis.

Assume, for this specification of the ARA, that the Owner believes that the Pirates are expected utility maximizers who derive the Owner's uncertainty about the Pirates' decision on whether to attack from her uncertainty about the Pirates' probabilities and utilities. Thus, the Owner must analyze the decision problem from the Pirates' perspective, as shown in Figure 4.7. Note that the set of alternatives for the Pirates has expanded to include alternatives $a^i \in \mathcal{A}$, for $i = 2, \ldots, n$, representing the Pirates' option to attack ships owned by others. Also, there are new chance nodes D_2 at the end of the tree paths starting at a^i, which represent the response of ships $i = 2, \ldots, n$ to a hijacking attempt, since these are uncertainties from the Pirates' standpoint. The Owner's analysis of the Pirates' decision making enables probabilistic assessment of the perceived preferences of the Pirates. Her uncertainty over these preferences is modeled through the random utility function $U_A(a, s, d_2)$, for $a \in \mathcal{A} = \{a^0, a^1, \ldots, a^n\}$. Her uncertainty about the Pirates' beliefs regarding a successful attack on her ship and the subsequent payoff is modeled by the random variables $P_A(S = 1 \mid a^1, d_1)$ and $P_A(D_2 \mid d_1, a^1, S = 1)$. And, similarly, her uncertainty about the Pirates' beliefs regarding attacks upon other ships is modeled by $P_A(S = 1 \mid a^i)$ and $P_A(D_2 \mid a^i, S = 1)$ for $i = 2, \ldots, n$. To begin, we assume that the n ships are similar.

The Owner thinks that the relevant outcomes for the Pirates are the net assets gained and the number of lives lost. Carney (2009) reports that the average cost of an attack is about €30,000 and the average ransom paid is about €2.3M. For simplicity, the Owner assumes that two pirates are killed when an attack is repelled but no pirates die in a successful attack. However, if the Owner calls in the Navy, then five pirates die. (As usual, it is straightforward to treat these outcomes as random variables, through a slightly more elaborate analysis.)

The Owner knows that her ship is less valuable to the Pirates than it is to her. The Pirates can sell its cargo, sell it for scrap, or use it as a mother ship for new attacks. So the Owner assesses the economic value of the ship to the Pirates at €1M. Also, suppose the Owner thinks that Pirates value the life of one of their own at €0.25M. Under these assumptions, Table 4.3 summarizes the payoffs to the Pirates under various attack scenarios. The aggregated monetary equivalent is c_A, in the last column.

The Owner also needs to model her beliefs about the extent to which Pirates are risk-seeking with respect to profits. Suppose she thinks they are constant absolute

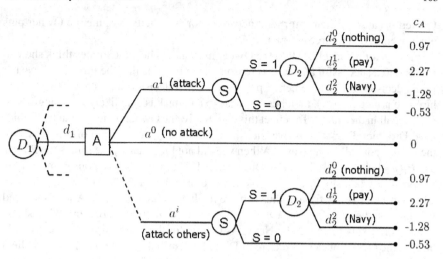

Fig. 4.7 Game tree representing the Owner's perception of the decision problem of the Somali pirates.

Table 4.3 Payoffs to the Pirates according to each paths in their game tree, $i = 1,\dots,n$. These calculations assume that all ships are equivalent in expectation for all relevant risks and rewards.

A	S	D_2	Ship kept	Profit	Lives lost	c_A
no attack			0	0	0	0.00
attack	$S=1$	don't pay	1	−0.03M	0	0.97
attack	$S=1$	pay ransom	0	2.27M	0	2.27
attack	$S=1$	call Navy	0	−0.03M	5	−1.28
attack	$S=0$		0	−0.03M	2	−0.53

risk seeking, and thus have utility function of the form $u_A(c_A) = \exp(c \times c_A)$, with $c > 0$. The Owner is not sure about the value c and assigns it the uniform distribution on $[0, 20]$. The uncertainty over c induces uncertainty over u_A, providing U_A.

The Pirates have beliefs about the probability of an attack being successful, conditional on the observed defensive choices made by the Owner. In order to model these, the Owner uses subjective distributions. Specifically, suppose she assumes that $P_A(S = 1 \mid a^1, d_1^0) \sim \text{Beta}(40,60)$ when no defensive action taken, that $P_A(S = 1 \mid a^1, d_1^1) \sim \text{Beta}(10,90)$ when one armed guard is hired, and that $P_A(S = 1 \mid a^1, d_1^2) \sim \text{Beta}(50,950)$ when two guards are hired. Here, the expected values of these distributions equal the assessed probabilistic beliefs of the Defender for the same situation, reflecting their common knowledge of past exploits, but in other situations the means could be different.

For the other ships, the Owner does not know whether they carry armed guards, nor, if they do, whether there are one or two of them. So, to her, the probability that the Pirates assign to a successful attack on another ship is a mixture of the three previous beta distributions. Of course, the Pirates have better information than the

Owner, because of their informant network. For specificity, assume the Owner puts equal weight on all three components.

Finally, for the level-2 ARA, the Owner must assess how the Pirates think she will respond to a successful attack. It is reasonable to imagine that the Pirates believe her initial decision is a clue to her response. If her first decision were aggressive (i.e., to hire two armed guards), then her response to an attack is also likely to be aggressive (i.e., to call in the Navy). To reflect this concretely, the Owner believes that the Pirates have Dirichlet distributions over her three response options (write off the loss, pay the ransom, or call in the Navy). When she did not hire guards, she thinks the Pirates use $P_A(D_2 \mid d_1^0, A = a^1, S = 1) \sim \text{Dirichlet}(1,1,1)$ (so all three responses are equally likely). But when she hired one guard, she thinks they use $P_A(D_2 \mid d_1^1, A = a^1, S = 1) \sim \text{Dirichlet}(0.1,4,6)$, and thus is more likely to call in the Navy. And if she hired two guards, then $P_A(D_2 \mid d_1^2, A = a^1, S = 1) \sim \text{Dirichlet}(0.1,1,10)$, implying she is much more likely to call in the Navy. As before, for the other ships that are possible targets, the Owner places a mixture of Dirichlet distributions over the Pirates' beliefs about their responses. It is convenient to assume that $P_A(D_2 \mid A = a^i, S = 1)$ puts equal weight on all three components.

With all this machinery, it is now possible for the Owner to find the level-1 solution. She uses backwards induction to solve her perception of the Pirates' decision problem in Fig. 4.7.

1. She computes the random expected utilities corresponding to the Pirates' selection of a^1, conditional on her initial defense choices $d_1 \in \mathcal{D}_1 \setminus \{d_1^3\}$:

$$\Psi_A(d_1, a^1) = \left[\sum_{d_2 \in \mathcal{D}_2} U_A(a^1, S = 1, d_2) \, P_A(D_2 = d_2 \mid d_1, a^1, S = 1) \right] \times$$
$$P_A(S = 1 \mid d_1, a^1) + P_A(S = 0 \mid d_1, a^1) \, U_A(a^1, S = 0).$$

2. She computes the random expected utilities corresponding to the Pirates' selection of a^i for $i = 2, \dots, n$:

$$\Psi_A(a^i) = \left[\sum_{d_2 \in \mathcal{D}_2} U_A(a^i, S = 1, d_2) \, P_A(D_2 = d_2 \mid a^i, S = 1) \right] \times$$
$$P_A(S = 1 \mid a^i) + P_A(S = 0 \mid a^i) \, U_A(a^i, S = 0).$$

3. She computes the Owner's predictive probabilities of being attacked ($A = a^1$) conditional on the initial defense choices, $d_1 \in \mathcal{D}_1 \setminus \{d_1^3\}$:

$$p_D(A = a^1 \mid d_1) = \Pr[\, \Psi_A(d_1, a^1) > \max\{U_A(a^0), \Psi_A(a^2), \dots, \Psi_A(a^n)\} \,].$$

These probabilities can be approximated through Monte Carlo simulation by drawing a sample $\{(u_A^k, p_A^k)\}_{k=1}^N \sim (U_A, P_A)$ from the Pirates' random utilities and probabilities, as assessed by the Owner, and then solving the Pirates' decision problem for each draw. This generates an estimate of the probability that the Owner's ship is chosen as the target. Specifically, with n possible ships to attack and N Monte

Carlo draws, the estimate is

$$\frac{\#\{1 \leq k \leq N : \psi_A^k(d_1,a^1) > \max\{u_A^k(a^0),\ \psi_A^k(a^2),\ldots,\psi_A^k(a^n)\}\}}{N}.$$

To illustrate, suppose there are nine possible ships that could be attacked. Under the modeling assumptions described, 50,000 Monte Carlo draws finds that the estimated probability that the Owner's ship will be attacked is $\hat{p}_D(A = a^1 \mid d_1^0) = 0.303$ if she does not hire guards, $\hat{p}_D(A = a^1 \mid d_1^1) = 0.026$ if she hires one armed guard, and $\hat{p}_D(A = a^1 \mid d_1^2) = 0.00004$ if she hires two armed guards. So the Owner's estimate of the probability of attack decreases rapidly when she obtains protection.

Given the estimated values of her beliefs about the attack probabilities, she solves her problem using backwards induction on the tree in Fig. 4.6. At decision node D_2, she finds her maximum utility action conditional on each $d_1 \in \mathcal{D}_1 \setminus \{d_1^3\}$:

$$d_2^*(d_1,a^1,S=1) = \operatorname{argmax}_{d_2 \in \mathcal{D}_2} u_D(c_D(d_1,S=1,d_2)).$$

Next, at chance node ⓢ she obtains her expected utilities as

$$\psi_D(d_1,a^1) = p_D(S=1 \mid d_1,a^1)\, u_D(c_D(d_1,S=1,d_2^*(d_1,a^1,S=1))) +$$
$$p_D(S=0 \mid d_1,a^1)\, u_D(c_D(d_1,S=0)).$$

At this point, she uses her assessments of the estimated probability of being attacked, conditional on her initial defense decision, or $\hat{p}_D(A = a^1 \mid d_1)$, to compute for each $d_1 \in \mathcal{D}_1 \setminus \{d_1^3\}$ her expected utility at chance node Ⓐ :

$$\psi_D(d_1) = \psi_D(d_1,a^1)\, \hat{p}_D(A = a^1 \mid d_1) + u_D(c_D(d_1,S=0))\, (1 - \hat{p}_D(A = a^1 \mid d_1)).$$

Finally, she finds her maximum expected utility decision at decision node $\boxed{D_1}$ as

$$d_1^* = \operatorname{argmax}_{d_1^i \in \mathcal{D}_1} \psi_D(d_1^i),$$

where $\psi_D(d_1^3) = u_D(c_D(d_1^3))$ is obtained from Table 4.2. The Defender's best strategy is to first choose d_1^* at node $\boxed{D_1}$, and, if the ship is hijacked, respond by choosing $d_2^*(d_1^*,a^1,S=1)$ at node $\boxed{D_2}$.

This case study made many assumptions. A sensitivity analysis can show their impact. First, recall that the Owner has a constant level of risk aversion, with utility function of the form $u_D(c_D) = -\exp(c \times c_D)$ for $c > 0$. Then, if there are nine ships that may be targeted by Pirates, the Owner's maximum expected utility choice for $c = 0.1$ and $c = 0.4$ is to hire one armed guard. For $c = 1$ and $c = 2$, she should hire two armed guards. In both cases, if hijacked, she should then pay the ransom. But for $c = 5$, she should take the Cape of Good Hope route.

To also explore the sensitivity of the Owner's optimal decision to assumptions about (a) the probability that an attack is successful given the initial defense decision, (b) the number of nearby ships that could be targeted, and (c) the probability of an attempted attack given the initial defense decision, consider Table 4.4. It shows the Owner's optimal initial and secondary decisions. If hijacked, it is always best for the Owner to pay the ransom. And, as the number of ships increases, risk diminishes and less aggressive initial choices are made. Similarly, as the estimates of attack probabilities diminish, or as the probability of successful attack diminishes, the Owner should choose less expensive initial defenses.

Table 4.4 A sensitivity analysis of the decision theory in the Somali pirates case study. Factors considered are different levels of risk aversion, different probabilities for successful attack, different numbers of nearby ships, and different probabilities of a successful attack given the initial decision.

$p_D(S=1\|a^1,d_1)$		n	$\hat{p}_D(A=a^1\|d_1)$			c	d^*	$d_2^*(d_1^*)$
d_1^1	d_1^2		d_1^0	d_1^1	d_1^2			
0.2	0.1	5	0.41010	0.18354	0.00382	0.1 – 1	d_1^2 (team)	d_2^1 (pay)
						2 – 5	d_1^3 (GH route)	
		9	0.27260	0.05780	0.00008	0.1	d_1^1 (man)	d_2^1 (pay)
						0.4 – 1	d_1^2 (team)	d_2^1 (pay)
						2 – 5	d_1^3 (GH route)	
		15	0.18322	0.01622	0.00000	0.1 – 0.4	d_1^1 (man)	d_2^1 (pay)
						1 – 5	d_1^2 (team)	d_2^1 (pay)
		20	0.14230	0.00628	0.00000	0.1 – 1	d_1^1 (man)	d_2^1 (pay)
						2 – 5	d_1^2 (team)	d_2^1 (pay)
0.10	0.05	5	0.46564	0.12166	0.00328	0.1	d_1^1 (man)	d_2^1 (pay)
						0.4 – 1	d_1^2 (team)	d_2^1 (pay)
						2 – 5	d_1^3 (GH route)	
		9	0.30332	0.02560	0.00004	0.1 – 0.4	d_1^1 (man)	d_2^1 (pay)
						1 – 2	d_1^2 (team)	d_2^1 (pay)
						5	d_1^3 (GH route)	
		15	0.19386	0.00392	0.00000	0.1 – 1	d_1^1 (man)	d_2^1 (pay)
						2 – 5	d_1^2 (team)	d_2^1 (pay)
		20	0.14836	0.00098	0.00000	0.1 – 1	d_1^1 (man)	d_2^1 (pay)
						2 – 5	d_1^2 (team)	d_2^1 (pay)
0.05	0.025	5	0.49010	0.09372	0.00374	0.1	d_1^1 (man)	d_2^1 (pay)
						0.4 – 1	d_1^2 (team)	d_2^1 (pay)
						2 – 5	d_1^3 (GH route)	
		9	0.31764	0.01596	0.00002	0.1 – 0.4	d_1^1 (man)	d_2^1 (pay)
						1 – 2	d_1^2 (team)	d_2^1 (pay)
						5	d_1^3 (GH route)	
		15	0.19842	0.00142	0.00000	0.1 – 1	d_1^1 (man)	d_2^1 (pay)
						2 – 5	d_1^2 (team)	d_2^1 (pay)
		20	0.14778	0.00024	0.00000	0.1 – 1	d_1^1 (man)	d_2^1 (pay)
						2 – 5	d_1^2 (team)	d_2^1 (pay)

4.4 Case Study: *La Relance*

To illustrate level-k thinking in ARA for a sequential game, we consider the famous gamble *La Relance*, which is played as follows:

> *La Relance*: Bart and Lisa each ante one dollar to play. Then each independently draws a private number from the uniform distribution on [0, 1]. Bart must either fold or bet a fixed amount b. If Bart bets, then Lisa either folds or calls (i.e., she wagers the amount b). If Lisa calls, then whomever drew the larger number wins the pot.

This simple game has attracted the attention of mathematicians, economists, and statisticians. Emile Borel (1938) found the minimax solution, anticipating modern game theory. von Neumann and Morgenstern (1944) extended Borel's result to allow players to check in multiple rounds (i.e., to pass without betting if no previous player has bet); Bellman and Blackwell (1949) examined the role of bluffing; and Karlin and Restrepo (1957) solved the problem of multiple players with multiple rounds and multiple increments of bet size. A recent account, with modern proofs and language, is given in Ferguson and Ferguson (2003).

Our analysis derives the ARA solution from Bart's perspective, under a variety of scenarios. Note that in *La Relance*, the sequence of play advantages Lisa. She can use the information from Bart's decision to improve her expected return.

Assume Bart observes that his draw is $X = x$. Lisa's draw $Y = y$ is unknown to him, but he knows it is Unif[0, 1]. Let V_x be the random amount won by Bart when he draws $X = x$. Table 1 shows the possible situations, depending on which players decide to bet and the values of X and Y.

V_x	Bart	Lisa	Outcome
-1	fold		
1	bet	fold	
$1+b$	bet	bet	$X > Y$
$-(1+b)$	bet	bet	$X < Y$

The *La Relance* problem can also be formulated as a game tree, shown in Fig. 4.8, which emphasizes the sequential aspect of the game.

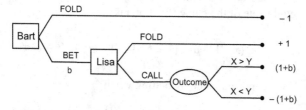

Fig. 4.8 Game tree representation of *La Relance*. Only Bart's payoff is represented. The oval represents the chance node when the players compare the numbers they drew.

From Table 1, and neglecting ties (an event of measure zero), the expected amount won by Bart, given his draw, is

$$\mathbb{E}[V_x] = -\mathbb{P}[\text{ Bart folds }] + \mathbb{P}[\text{ Bart bets and Lisa folds }] +$$
$$(1+b)\mathbb{P}[\text{ Lisa calls and loses }] - (1+b)\mathbb{P}[\text{ Lisa calls and wins }].$$

Bart seeks the decision that maximizes $\mathbb{E}[V_x]$, and must find his optimal "betting function" $g(x)$; given $X = x$, he bets with probability $g(x)$ (which may take the value 0 or 1). Then

$$\mathbb{E}[V_x] = -[1 - g(x)] + g(x)\mathbb{P}[\text{ Lisa folds } | \text{ Bart bets }] + \quad (4.3)$$
$$(1+b)g(x)\mathbb{P}[\text{ Bart wins } | \text{ Lisa calls }]\mathbb{P}[\text{ Lisa calls } | \text{ Bart bets }] -$$
$$(1+b)g(x)\mathbb{P}[\text{ Bart loses } | \text{ Lisa calls }]\mathbb{P}[\text{ Lisa calls } | \text{ Bart bets }].$$

So, to derive his optimal play, Bart needs to know both $\mathbb{P}[\text{ Lisa calls } | \text{ Bart bets }]$ and $\mathbb{P}[\text{ Bart wins } | \text{ Lisa calls }]$.

This ARA follows Banks, Petralia and Wang (2011). The first step is for Bart to build a model for Lisa's strategic analysis. If he is a level-1 thinker, he supposes that she is a level-0 thinker. Thus Lisa is not strategic, but her non-strategic thinking could take several forms. For example, she might bet at random, without regard to the number she drew. Alternatively, if she has drawn the number y, she might bet with probability y. Or, her rule might be to bet if $y > c$ and fold if $y \leq c$, where c is some number she chooses without consideration of Bart's analysis.

If Bart thinks Lisa bets when $y > c$ and otherwise folds, then he should place a subjective distribution $\pi(c)$ on Lisa's threshold value c. Given $\pi(c)$, Bart's thinks the probability that Lisa will bet is

$$\mathbb{P}[\text{ Lisa bets }] = \int_0^1 \int_c^1 \pi(c)\,dy\,dc = \int (1-c)\pi(c)\,dc = 1 - \mu,$$

where μ is the mean of $\pi(c)$. Calculation shows

$$\mathbb{P}[\text{ Bart wins with } x \mid \text{ Lisa calls }] = \int_0^x \int_c^x \pi(c)\,dy\,dc = \int_0^x (x-c)\pi(c)\,dc.$$

When $X = x$, suppose Bart bets with probability $g(x)$. Set $\gamma = (1+b)(1-\mu)$. Using (4.3), the expected value of the game (for Bart) is

$$\mathbb{E}[V_x] = -1 + g(x)\left[1 + \mu - (1+b) + 2(1+b)\int_0^x (x-c)\pi(c)\,dc\right].$$

Bart maximizes his expected value by betting when

$$m(x) = 1 + \mu - (1+b) + 2(1+b)\int_0^x (x-c)\pi(c)\,dc > 0$$

and folding when $m(x) < 0$; he may do as he pleases when $m(x) = 0$.

Example 4.1: Suppose that Bart thinks $\pi(c)$ has the Beta(α, β) distribution. Then

$$\int_0^x (x-c)\pi(c)\,dc = xI_x(\alpha, \beta) - \frac{\alpha}{\alpha+\beta}I_x(\alpha+1, \beta) \qquad (4.4)$$

where

$$I_x(\alpha, \beta) = \frac{\Gamma(\alpha+\beta)}{\Gamma(\alpha)\Gamma(\beta)}\int_0^x t^{\alpha-1}(1-t)^{\beta-1}\,dt$$

is the regularized incomplete beta function. Consider, for example, three cases: (1) $\alpha = \beta = 1$, in which Bart is completely agnostic about the location of the step in Lisa's betting function; (2) $\alpha = \beta = 2$, so Bart assumes that Lisa's step is near $y = 1/2$; and (3) $\alpha = 3, \beta = 1$, so Bart assumes that Lisa is conservative, calling when Y is relatively large. For the first case, equation (4.4) reduces to $\frac{1}{2}x^2$. Bart bets if and only if

$$x > \left(1 - \frac{3}{1+b}\right)^{1/2}.$$

For the second case, (4.4) equals $x^3 - \frac{1}{2}x^4$ and Bart bets if and only if

$$x^3 - \frac{1}{2}x^4 > \frac{1}{2}\left(1 - \frac{3}{1+b}\right).$$

In the third case, (4.4) is $\frac{1}{4}x^4$ and Bart bets if and only if

$$x > \left(2 - \frac{14}{1+b}\right)^{1/4}.$$

The monotonicity in b implies that as the bet size increases, Bart should become more conservative.

Now suppose Bart is a level-2 thinker, so he assumes that Lisa is a level-1 thinker, and he must model Lisa's reasoning in order to develop his belief about her decision rule. Bart accomplishes this by doing the analysis he expects her to make, using subjective distributions to describe the quantities he does not know.

Bart knows that Lisa's opinion about his X is updated by his decision to bet. And suppose Bart believes that Lisa thinks his betting function is $\tilde{g}(x)$. So Bart believes that Lisa calculates her conditional density of X, given that Bart chose to bet, as $\tilde{f}(x) = \tilde{g}(x)/\int \tilde{g}(w)\,dw$. Note that if Bart believes Lisa thinks his betting rule is to fold when x is less than some value x_0 but he always bet when it is greater, then $\tilde{g}(x)$ is a step function and her posterior distribution on X would be truncated below x_0 with the weight reallocated proportionally to values above x_0.

Of course, Bart will have uncertainty about $\tilde{g}(x)$; e.g., he might think that there probability 0.5 that Lisa thinks $\tilde{g}(x) = x$, and probability 0.5 that he bets only when $x > b/(b+2)$. In such situations, Bart the Bayesian should have a subjective distribution $\pi(g)$ over the set $\mathcal{G} = \{g : 0 \le g(x) \le 1 \text{ for all } x \text{ such that } 0 \le x \le 1 \in\}$ of all possible betting functions, and he would use the $\tilde{g}(x)$ that satisfies

$$\tilde{g}(x) = \int_{\mathcal{G}} g(x)\pi(g)\,dg.$$

This solution is impractical, but Bart could approximate this $\tilde{g}(x)$ by mixing over a small number of distinct betting functions.

Given $\tilde{g}(x)$ and the corresponding $\tilde{f}(x)$, Bart believes that Lisa thinks her probability of winning, conditional on Bart's bet, is

$$\mathbb{P}[X \le y \mid \text{Bart bets}] = \tilde{F}(y) = \int_0^y \tilde{f}(z)\,dz,$$

where y is unknown to Bart. And Bart believes that Lisa will call when the expected value of her winnings V_y from betting b is greater than one dollar, the amount she loses by folding; i.e., Lisa calls if

$$(1+b)\tilde{F}(y) - (1+b)[1 - \tilde{F}(y)] \ge -1.$$

Solving shows that Bart thinks Lisa will fold if $\tilde{F}(y) < b/[2(1+b)]$, call if $\tilde{F}(y) > b/[2(1+b)]$, and do as she likes when $\tilde{F}(y) = b/[2(1+b)]$.

Let $\tilde{y} = \inf\{y : \tilde{F}(y) > b/[2(1+b)]\}$. Lisa draws $Y > \tilde{y}$ with probability $1 - \tilde{y}$ and this is Bart's best guess about the chance that she calls. When Lisa draws a number larger than $\tilde{y}$, then $\mathbb{P}[\text{Bart wins} \mid \text{Lisa calls}] = [(x-\tilde{y})/(1-\tilde{y})]^+$, where $[\cdot]^+$ is zero if the argument is negative.

Therefore, Bart believes his expected value of the game is:

$$\mathbb{E}[V_x] = -[1 - g(x)] + g(x)\tilde{y} + (1+b)g(x)[x-\tilde{y}]^+ - (1+b)g(x)(1-\tilde{y}-[x-\tilde{y}]^+)$$

or

$$\mathbb{E}[V_x] = \begin{cases} -1 + g(x)(2\tilde{y} + b\tilde{y} - b) & \text{if } x \le \tilde{y} \\ -1 + g(x)(2x + 2bx - b\tilde{y} - b) & \text{if } x > \tilde{y}. \end{cases} \tag{4.5}$$

Bart must choose $g(x)$ to maximize $\mathbb{E}[V_x]$.

Let $\gamma = b/(b+2)$. When $x \le \tilde{y}$, Bart should bet if $\tilde{y} > \gamma$, fold if $\tilde{y} < \gamma$ and do as he likes when $\tilde{y} = \gamma$. When $x > \tilde{y}$, then, for

$$\tilde{x} = \frac{b(1+\tilde{y})}{[2(1+b)]},$$

Bart should bet when $x > \tilde{x}$, fold when $x < \tilde{x}$, and do as he likes when $x = \tilde{x}$. So there are three cases, depending on whether Bart thinks Lisa's $\tilde{y}$ is equal to, less than, or greater than γ. Equality corresponds to Lisa using the minimax rule. If $\tilde{y} < \gamma$, then Lisa is rash, and if $\tilde{y} < \gamma$, then Lisa is too conservative.

Case I: Bart Believes that Lisa Plays Minimax

Lisa's traditional minimax solution has $\tilde{y} = \gamma$ (von Neumann and Morgenstern, 1944, Chap. 9). In that case, Bart's minimax solution is to always bet if $x > \tilde{y}$, and bet with probability $2/(b+2)$ when $x \leq \tilde{y}$. The expected value of the game to Bart is $-\gamma^2$, which agrees with our intuition that Bart's is disadvantaged by betting first, which implies the negative expected value.

In contrast, ARA shows that when Lisa uses the minimax threshold $\tilde{y} = \gamma$, then Bart should bet if $x > \gamma$, as usual. But he may bet or not, as he chooses, when $x \leq \gamma$. This is a bit different from the minimax solution. The discrepancy arises because, if Lisa knows that Bart's is not betting with probability $2/(b+2)$ when $x \leq \gamma$, she can then increase her expected payoff by changing the threshold at which she calls. For example, suppose Bart bets if and only if $x > \gamma$. Then

$$g(x) = \begin{cases} 0 & \text{if } 0 \leq x \leq \gamma \\ 1 & \text{if } \gamma < x \leq 1. \end{cases}$$

If Lisa thought this was Bart's rule, she would calculate

$$\tilde{F}(y) = \begin{cases} 0 & \text{if } 0 \leq y \leq \gamma \\ \frac{y-\gamma}{1-\gamma} & \text{if } \gamma < x \leq 1 \end{cases}$$

and invert it to find

$$\tilde{y} = \tilde{F}^{-1}\left(\frac{b}{2(b+1)}\right) = \frac{b}{b+1}.$$

So Lisa should not call when $y > \gamma$. She could increase the expected value of her game by raising the threshold at which she calls to $b/(b+1)$.

Now suppose Bart uses a minimax rule. This need not contradict the ARA solution. Specifically, Bart could play the admissible rule of Ferguson and Ferguson (2003):

$$g(x) = \begin{cases} 0 & \text{if } 0 \leq x \leq \gamma^2 \\ 1 & \text{if } \gamma^2 < x \leq 1. \end{cases}$$

If Lisa knew that Bart used this betting function, then she would calculate

$$\tilde{F}(y) = \begin{cases} 0 & \text{if } 0 \leq y \leq \gamma^2 \\ \frac{y-\gamma^2}{1-\gamma^2} & \text{if } \gamma^2 < x \leq 1 \end{cases}$$

and invert this equation to find

$$\tilde{y} = \tilde{F}^{-1}\left(\frac{b}{2(b+1)}\right) = \gamma.$$

Thus, when Bart plays an ARA rule that is also a minimax rule, Lisa's optimal threshold for betting is γ.

Case II: Bart Believes that Lisa Is Risk Seeking

Suppose Bart believes that Lisa is reckless, calling with $\tilde{y} < \gamma$. The preceding ARA shows that Bart's betting function should be

$$g(x) = \begin{cases} 0 & \text{if } 0 \leq x \leq \max\{\tilde{y},\tilde{x}\} \\ 1 & \text{if } \max\{\tilde{y},\tilde{x}\} < x \leq 1, \end{cases}$$

where $\tilde{x} = [b(1+\tilde{y})]/[2(1+b)]$. Algebra shows that if $\tilde{y} < \gamma$, then $\tilde{x} > \tilde{y}$.

The value of this game for Bart is

$$\int \mathbb{E}[V_x]\,dx = -\int_0^{\tilde{x}} dx + \int_{\tilde{x}}^1 (-1+2x+2bx-b\tilde{y}-b)\,dx$$

$$= b\tilde{x} - b\tilde{y}(1-\tilde{x}) - (1+b)\tilde{x}^2.$$

This value is strictly larger than $-\gamma^2$, Bart's minimax value. For example, when Lisa always calls (i.e., $\tilde{y} = 0$), then Bart's value for the game is $b^2/(4+4b)$. So, if Bart correctly believes that Lisa is reckless, then his game can have positive value, despite her second-mover advantage.

Case III: Bart Believes that Lisa Is Risk Averse

Suppose that Bart thinks that Lisa is conservative, calling when $\tilde{y} > \gamma$. This implies that $\tilde{x} < \tilde{y}$. When $x > \tilde{y}$, then

$$\mathbb{E}[V_x] = -(1-g(x)) + g(x)\tilde{y} + (1+b)g(x)(1-\tilde{y})\frac{x-\tilde{y}}{1-\tilde{y}}$$

$$-(1+b)g(x)(1-\tilde{y})\left(1 - \frac{x-\tilde{y}}{1-\tilde{y}}\right).$$

When $x > \tilde{y}$, Bart should always bet. And when $x < \tilde{y}$, Bart's expected payoff is

$$\mathbb{E}[V_x] = -1 + g(x)\left[1 + \tilde{y} - (1+b)(1-\tilde{y})\right]. \tag{4.6}$$

For $\tilde{y} > \gamma$, the quantity within the brackets is strictly positive. Thus, when $x < \tilde{y}$, the optimal $g(x)$ is equal to 1: Bart should always bet.

Bart's value for this game is

$$\int_0^1 \mathbb{E}[V_x]\,dx = \int_0^{\tilde{y}} [\tilde{y} - (1+b)(1-\tilde{y})]\,dx$$

$$+ \int_{\tilde{y}}^1 [\tilde{y} + (1+b)(x-\tilde{y}) - (1+b)(1-x)]\,dx$$

$$= -b\tilde{y} + \tilde{y}^2(1+b).$$

This value increases in $\tilde{y}$ for $\tilde{y} > \gamma$ and is equal to the minimax value at $\tilde{y} = \gamma$. Thus Bart's value of the ARA game when Lisa is conservative is strictly larger than the minimax value.

This level-k ARA can be extended. For $k = 3$, Bart would assess the betting function that he believes Lisa thinks Bart is using to describe her guess about Bart's betting function. Then his solution for Lisa's strategy must take account of the fact that Lisa is modeling Bart's thinking in order to develop her own betting rule. The analysis is straightforward but intricate. It is more productive to examine the ways in which the ARA perspective enables solutions of generalized versions of *La Relance*.

4.4.1 Continuous Bets

Suppose Bart may make a continuous wager. Instead of being having to either bet b dollars or fold, he may place any positive bet he wishes, in some known interval $[\varepsilon, K]$, where $0 < \varepsilon < K < \infty$. This is an difficult problem for minimax analysis; Karlin and Restrepo (1957) finds the solution when there is a fixed number of distinct bids that are allowed, and Newman (1959) obtains a full solution. Ferguson, Ferguson and Gawargy (2007) mentions unpublished work by W. H. Cutler in 1976 on continuous bets in the context of endgame poker. Banks, Petralia and Wang (2011) addresses the problem from an ARA perspective.

Consider the ARA perspective on continuous bet sizes, using the level-2 solution concept. Define the following notation:

$g(x)$: the probability that Bart chooses to bet, given $X = x$.

$h(b|x)$: a probability density on $[\varepsilon, K]$ that Bart uses to select his bet, given that he decides to bet.

B_x: the random variable with value in $[\varepsilon, K]$ that is Bart's bet after he learns $X = x$.

Let $\mathbb{P}_{h(\cdot|x)}[\cdot]$ and $\mathbb{E}_{h(\cdot|x)}[\cdot]$ denote the probability and expectation induced by the density $h(\cdot|x)$, respectively.

In a level-2 ARA, Bart models Lisa's opinion about his value of X given that she observes Bart's bet $B_x = b$. So define

$\tilde{g}(x)$: Bart's belief about Lisa's belief about the probability that he decides to bet given $X = x$.

$\tilde{h}(b|x)$: Bart's belief about Lisa's belief about the density function on $[\varepsilon, K]$ that Bart will use to select his bet given that he decides to bet.

$\tilde{f}(x|b)$: Bart's belief about Lisa's posterior density for X after she observes that Bart bets the amount b; specifically, a coherent Lisa must think

$$\tilde{f}(x|b) = \frac{\tilde{h}(b|x)\tilde{g}(x)}{\int_0^1 \tilde{h}(b|z)\tilde{g}(z)\,dz}.$$

In this framework, Bart's expected payoff given $X = x$ and the strategy pair $g(x), h(\cdot|x)$ is

$$\mathbb{E}\left[V_B | X = x\right] = g(x)\left\{\mathbb{E}_{h(\cdot|x)}\left[\mathbb{P}_{\tilde{f}(\cdot|B_x)}[\,\text{Lisa folds} \mid \text{Bart bets } B_x] \mid X = x\right] + \right. \quad (4.7)$$

$$\mathbb{E}_{h(\cdot|x)}\left[\mathbb{P}_{\tilde{f}(\cdot|B_x)}[\,\text{Lisa loses} \mid \text{Bart bets } B_x]\cdot(1+B_x) \mid X = x\right] -$$

$$\left.\mathbb{E}_{h(\cdot|x)}\left[\mathbb{P}_{\tilde{f}(\cdot|B_x)}[\,\text{Lisa wins} \mid \text{Bart bets } B_x]\cdot(1+B_x) \mid X = x\right]\right\} -$$

$$\underbrace{(1-g(x))}_{\text{Bart folds}}.$$

So Bart's ARA solution, denoted by $\{g^*(x), h^*(\cdot|x)\}$, is

$$\{g^*(x), h^*(\cdot \mid x)\} \in \underset{g(x),h(\cdot|x)}{\text{argmax}} \ \ \mathbb{E}_{g(x),h(\cdot|x)}\left[V_B \mid X = x\right]. \quad (4.8)$$

In order to solve for $\{g^*(x), h^*(\cdot|x)\}$, Bart must study Lisa's strategy and then use backwards induction.

When Lisa folds, her payoff is -1. And if Bart bets B_x and if he believes that Lisa will form the posterior assessment $\tilde{f}(\cdot|b)$ for his X conditional on his bet, then Bart believes that Lisa's assessment of her probability of winning with $Y = y$ is

$$Pb_{\tilde{f}(\cdot|B_x)}[X \leq Y \mid B_x, Y = y] = \int_0^y \tilde{f}(z \mid B_x)\,dz.$$

So Bart believes that, when Lisa calls, she expects a payoff of

$$\begin{aligned}
\mathbb{E}[V_y] &= \mathbb{P}_{\tilde{f}(\cdot|B_x)}[\,\text{Lisa wins} \mid B_x, Y = y, \text{ Lisa calls }]\cdot(1+B_x) - \\
&\quad \mathbb{P}_{\tilde{f}(\cdot|B_x)}[\,\text{Lisa loses} \mid B_x, Y = y, \text{ Lisa calls }]\cdot(1+B_x) \\
&= \mathbb{P}_{\tilde{f}(\cdot|B_x)}[X \leq Y \mid B_x, Y = y]\cdot(1+B_x) - \\
&\quad \left\{1 - \mathbb{P}_{\tilde{f}(\cdot|B_x)}[X \leq Y \mid B_x, Y = y]\right\}\cdot(1+B_x) \\
&= 2\mathbb{P}_{\tilde{f}(\cdot|B_x)}[X \leq Y \mid B_x, Y = y]\cdot(1+B_x)\cdot(1+B_x) - (1+B_x) \\
&= 2(1+B_x)\int_0^y \tilde{f}(z \mid B_x)\,dz - (1+B_x).
\end{aligned}$$

Bart therefore believes that Lisa will call if and only if

$$-1 \leq 2(1+B_x)\int_0^y \tilde{f}(z \mid B_x)\,dz - (1+B_x).$$

Since $\tilde{f}(z \mid B_x) \geq 0$, then there exists some $\tilde{y}^*(B_x)$ such that for all $y \geq \tilde{y}^*(B_x)$ she must have

$$\int_0^y \tilde{f}(z \mid B_x)\,dz \geq \int_0^{\tilde{y}^*} (B_x)\tilde{f}(z \mid B_x)\,dz \geq \frac{B_x}{2(1+B_x)}.$$

Thus, Lisa should call if and only if

$$Y \geq \tilde{y}^*(B_x) \equiv \inf\left\{y \in [0,1] : \int_0^y \tilde{f}(z|B_x)\,dz \geq \frac{B_x}{2(1+B_x)}\right\}. \quad (4.9)$$

So Bart thinks that the probability that Lisa calls after he bets B_x is

$$\mathbb{P}_{\tilde{f}(\cdot|B_x)}[\text{ Lisa calls } | \text{ Bart bets } B_x] = \mathbb{P}[Y \geq \tilde{y}^*(B_x) | B_x] = 1 - \tilde{y}^*(B_x).$$

With this machinery, Bart can calculate the following quantities:

$$\mathbb{P}_{\tilde{f}(\cdot|B_x)}[\text{ Lisa folds } | \text{ Bart bets } B_x] = \tilde{y}^*(B_x)$$

$$\mathbb{P}_{\tilde{f}(\cdot|B_x)}[\text{ Lisa loses } | \text{ Bart bets } B_x] = \mathbb{P}[\tilde{y}^*(B_x) \leq Y \leq x | B_x]$$
$$= [x - \tilde{y}^*(B_x)]^+$$

$$\mathbb{P}_{\tilde{f}(\cdot|B_x)}[\text{ Lisa wins } | \text{ Bart bets } B_x] = \mathbb{P}_{\tilde{f}(\cdot|B_x)}[\text{ Lisa calls } | \text{ Bart bets } B_x] -$$
$$\mathbb{P}_{\tilde{f}(\cdot|B_x)}[\text{ Lisa loses } | \text{ Bart bets } B_x]$$
$$= 1 - \tilde{y}^*(B_x) - [x - \tilde{y}^*(B_x)]^+.$$

Plugging these into (4.7) gives

$$\mathbb{E}_{f(x),g(\cdot|x)}[V_B | X = x] = -[1 - g(x)] +$$
$$g(x)\mathbb{E}_{h(\cdot|x)}\left[\tilde{y}^*(B_x) + 2[x - \tilde{y}^*(B_x)]^+(1 + B_x) - (1 - \tilde{y}^*(B_x))(1 + B_x)\right].$$

To finish, a technical result is needed.

Lemma 4.1. *If $\tilde{f}(\cdot|b)$ is positive and continuous in $b \in [\varepsilon, K]$, then $\tilde{y}^*(b)$ is continuous in b.*

Proof. Continuity and positivity of $\tilde{f}(\cdot | b)$ in b imply continuity of $\int_0^y \tilde{f}(z | b) dz$ in (y, b). The positivity of $\tilde{f}(\cdot | b)$ implies the (global) one-to-one condition specified in Jittorntrum (1978). Hence, $\tilde{y}^*(b)$ is the (unique) solution to

$$\int_0^y \tilde{f}(z | b) dz - \frac{b}{2(1 + b)} = 0, \quad b \in [\varepsilon, K],$$

and must be continuous in b. $\quad\square$

This lemma immediately provides the main result.

Theorem 4.1. *For all $x \in [0, 1]$, let $\tilde{f}(\cdot|b)$ be continuous and positive in $b \in [\varepsilon, K]$. Let $\tilde{y}^*(b)$ be defined as in Lemma 4.1. And let*

$$b^*(x) \in \underset{b \in [\varepsilon, K]}{\text{argmax}} \ \tilde{y}^*(b) + 2(x - \tilde{y}^*(b))^+(1 + b) - (1 - \tilde{y}^*(b))(1 + b),$$

$$\Delta^*(x) \equiv \underset{b \in [\varepsilon, K]}{\max} \ \tilde{y}^*(b) + 2(x - \tilde{y}^*(b))^+(1 + b) - (1 - \tilde{y}^*(b))(1 + b).$$

Then, Bart's level-2 ARA solution is

$$g^*(x) = \begin{cases} 0 & \text{if } \Delta^*(x) < -1 \\ 1 & \text{if } \Delta^*(x) \geq -1; \end{cases}$$
$$h^*(b \mid x) = \delta(b - b^*(x)),$$

where $\delta(\cdot)$ is the Dirac delta function.

Therefore, when Bart observes $X = x$, he will fold with probability 1 if $\Delta^*(x) < -1$, and bet $b^*(x)$ with probability 1 if $\Delta^*(x) \geq -1$. (The regularity condition that $\tilde{f}(\cdot \mid b)$ be positive and continuous in $b \in [\varepsilon, K]$ is sufficient but not necessary.)

Example 4.4: Assume that $\tilde{f}(\cdot \mid b)$ has the form

$$\tilde{f}(x|b) = \begin{cases} \frac{1+K}{1+b} & \text{if } 0 \leq x \leq \frac{1+b}{1+K} \\ 0 & \text{otherwise.} \end{cases} \tag{4.10}$$

Then, it is simple to show that $\tilde{y}^*(b) = \frac{b}{2(1+K)}$, and

$$\tilde{y}^*(b) + 2(x - \tilde{y}^*(b))^+(1+b) - (1 - \tilde{y}^*(b))(1+b)$$
$$= \begin{cases} -\frac{b^2}{2(1+K)} + (2x-1)(b+1) & \text{if } b \leq 2(1+K)x \\ \frac{b^2}{2(1+K)} - \frac{K}{1+K}b - 1 & \text{if } b > 2(1+K)x. \end{cases}$$

Suppose ε is sufficiently small that $\frac{\varepsilon^2 + 2(1+K)\varepsilon}{4(1+K)(1+\varepsilon)} < \frac{1}{2} + \frac{\varepsilon}{2(1+K)}$. Letting

$$A = \frac{\varepsilon^2 + 2(1+K)\varepsilon}{4(1+K)(1+\varepsilon)} \qquad B = \frac{1}{2} + \frac{\varepsilon}{2(1+K)} \qquad C = \frac{1}{2} + \frac{K}{2(1+K)}$$

one sees that there are four cases:

1. If $x < A$, then $b^*(x) = \varepsilon$ and $\Delta^*(x) = -\varepsilon^2/[2(1+K)] + (2x-1)(\varepsilon+1) < -1$. From Theorem 4.1, $g^*(x) = 0$; i.e., Bart must fold.
2. If $A \leq x < B$, then $b^*(x) = \varepsilon$ and $\Delta^*(x) = -\varepsilon^2/[2(1+K)] + (2x-1)(\varepsilon+1) \geq -1$. So $g^*(x) = 1$ and $h^*(b \mid x) = \delta(b - \varepsilon)$; i.e., Bart bets ε.
3. If $B \leq x < C$, then $b^*(x) = 2(1+K)x - (1+K)$ and $\Delta^*(x) = [(1+K)(2x-1)^2]/2 + (2x-1) \geq -1$. So $g^*(x) = 1$ and $h^*(b \mid x) = \delta(b - (2(1+K)x - (1+K)))$; i.e., Bart bets $2(1+K)x - (1+K)$.
4. If $x \geq C$, then $b^*(x) = K$ and $\Delta^*(x) = -K^2/[2(1+K)] + (2x-1)(K+1) \geq -1$. So $g^*(x) = 1$ and $h^*(b \mid x) = \delta(b - K)$; i.e., Bart bets K.

Figure 4.9 shows Bart's level-2 thinking ARA solution as a function of his draw $X = x$.

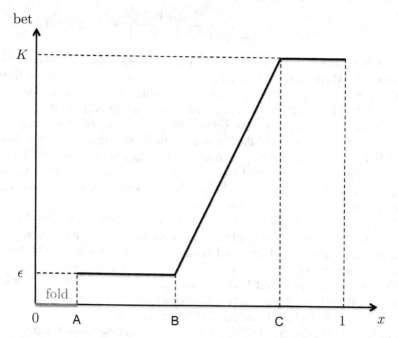

Fig. 4.9 This plot shows Bart's optimal decision for placing a continuous wager, as obtained from a level-2 ARA when Bart assumes that Lisa's posterior, given his bet, is uniform.

4.4.2 *Generalizations of* La Relance

The analysis of *La Relance* can be extended to the case in which the draws are not from a uniform distribution, and the case in which there are multiple opponents. Non-uniform draws are especially important when there is dependence, as would happen when drawing without replacement from a deck of cards. But it is technically challenging to find a minimax solution for such generalizations. Sakaguchi and Sakai (1981) gives some results for the Fairlie-Gumbel-Morgenstern distribution, in which the joint density has uniform marginals with a single parameter controlling dependence. Also, Sakaguchi (1984) and Karlin (1959, Chap. 4.3) consider special cases, such as independent but not identically distributed draws.

However, these generalizations are simple from the ARA perspective. First, suppose Bart and Lisa make independent draws from some continuous distribution $F(w)$. In that case, the conditional distribution that Bart imputes to Lisa is

$$\tilde{f}(x) = \frac{\tilde{g}(W(x))w(x)}{\int \tilde{g}(W(z))w(z)\,dz}$$

and Bart's betting function takes its step at

$$\tilde{x} = \frac{1}{2}\left[1 - \frac{1}{1+b}\frac{1+W(\tilde{y})}{1-W(\tilde{y})}\right],$$

which follows directly from the Probability Integral Transformation. The more useful extension, in which Bart and Lisa draw from a bivariate, possibly discrete distribution $W(x,y)$ (e.g., a deck of cards) is straightforward; Bart's distribution for Y is the conditional $W(y \mid X = x)$, and he knows that Lisa's analysis is symmetric.

For the case of multiple players, suppose that Bart, Lisa, and Milhouse are playing. Each antes one dollar, and then independently draws a uniform random number on $[0,1]$. Bart sees $X = x$ and may bet b units or fold. Lisa knows $Y = y$ and Bart's decision; she may then bet b or fold. Milhouse knows $Z = z$ and both Lisa's and Bart's decisions, and he may fold or bet b. Those players who did not fold then compare their numbers, and the highest number wins the pot.

The previous ARA applies as before, with the interesting extension that Bart must describe the opinions that Lisa and Milhouse have about each other, making it possible to model new kind of irrational behavior (e.g., Bart thinks that Milhouse is reckless with respect to Bart's signal but conservative with respect to Lisa's). Also, Bart can model Lisa as a level-1 thinker and Milhouse as a level-0 thinker. Kadane (2009, p. 241) points out that Edgar Allen Poe's Inspector Dupin tells of schoolboy who drew such distinctions in the game of "even-and-odd."

The level-1 ARA is straightforward. Table 4.5 shows the possible outcomes for Bart, conditional on $X = x$. He needs to find his betting function $g(x)$, which is the probability with which he should bet when he draws $X = x$. He has a subjective distribution for Lisa's betting function $g_L(y)$; he also has subjective distributions π_{10}, π_{11} for the thresholds c_{10}, c_{11} at which Milhouse will bet, when Bart bets but Lisa folds, and when both Bart and Lisa bet, respectively. Bart performs a Bayesian risk analysis to determine the strategy that maximizes his expected winnings, and that analysis assumes that neither Lisa nor Milhouse are strategic; i.e., neither attempts to model Bart's reasoning, nor each other's.

Table 4.5 Bart's possible payoffs when playing Lisa and Milhouse, according the decisions made by all parties and the outcomes of their draws.

V_x	Bart	Lisa	Milhouse	Outcome
-1	fold			
2	bet	fold	fold	
$2+b$	bet	bet	fold	$X > Y$
$2+b$	bet	fold	bet	$X > Z$
$2+2b$	bet	bet	bet	$X > Y,Z$
$-(1+b)$	bet	bet	fold	$X < Y$
$-(1+b)$	bet	fold	bet	$X < Z$
$-(1+b)$	bet	bet	bet	$X < Y$ or $X < Z$

For Bart's level-2 analysis, he assumes that both Lisa and Milhouse are level-1 thinkers. The expected value of the game to Bart is

$$\mathbb{E}[V_x] = -\mathbb{P}[\text{ Bart folds }] + 2\mathbb{P}[\text{ Bart bets and Lisa and Milhouse fold }] +$$
$$(2+b)\mathbb{P}[\text{ Lisa calls and loses and Milhouse folds }] +$$
$$(2+b)\mathbb{P}[\text{ Lisa folds and Milhouse calls and loses }] +$$
$$(2+2b)\mathbb{P}[\text{ all bet and Bart wins }] -$$
$$(1+b)\mathbb{P}[\text{ Lisa bets and wins, Milhouse folds }] -$$
$$(1+b)\mathbb{P}[\text{ Lisa folds, Milhouse bets and wins }] -$$
$$(1+b)\mathbb{P}[\text{ Lisa and Milhouse bet, Bart loses }].$$

The complexity of the problem requires a bit of notation. The function $g_{IJ}(x)$ is what Bart thinks player I thinks is the betting function for player J, and $g_{IJK}(x)$ is what Bart thinks player I thinks player J thinks is the betting function for player K. Generally, these will depend on preceding bets and folds, in which case the conditioning is denoted by the usual vertical stroke.

Bart's analysis starts with Milhouse, the last decision maker. Milhouse can observe four things: Both Bart and Lisa fold, both bet, Bart bets and Lisa folds, or Bart folds and Lisa bets. From Bart's perspective, he need not analyze the situations in which he folds—he knows his payoff will be -1.

First suppose that both Bart and Lisa bet. Milhouse should bet if his expected value for the game is greater than -1. Let B, L, and M denote Bart, Lisa, and Milhouse, respectively. Then Milhouse's probability of winning is

$$\mathbb{P}[\text{ M wins } | \text{ B, L bet }] = \mathbb{P}[X < z \mid \text{ B bets }]\mathbb{P}[Y < z \mid \text{ B, L bet }]$$

and the amount he could win is $2 + 2b$, the total bets of Bart and Lisa. If Bart believes that

- Milhouse thinks Bart's betting function is $g_{MB}(x)$, and
- Milhouse thinks Lisa's betting function, conditional on Bart having bet, is $g_{ML}(y \mid \text{ B bets })$

then

$$\mathbb{P}[\text{ M wins } | Z = z] = \mathbb{P}[X < z \mid \text{ B bets }]\mathbb{P}[Y < z \mid \text{ B, L bet }]$$
$$= \frac{\int_0^z g_{MB}(x)\,dx}{\int_0^1 g_{MB}(x)\,dx} \times \frac{\int_0^z g_{ML}(y \mid \text{ B bet })\,dy}{\int_0^1 g_{ML}(y \mid \text{ B bet })\,dy}.$$

Bart believes Milhouse will bet in order to maximize the value of his game, so Bart believes that Milhouse bets if and only if $Z > \tilde{z}_1$, where $\tilde{z}_1$ solves

$$2(1+b)\mathbb{P}[\text{ M wins } | Z = z] - (1+b)\mathbb{P}[\text{ M loses } | Z = z] = -1. \tag{4.11}$$

So Bart believes that Milhouse's betting function $g_M(z \mid B, L \text{ bet})$ must jump at $\tilde{z}_1$, the point at which Milhouse's expected winnings equal his loss from folding.

If Bart bets but Lisa folds, then similar arguments show that Bart believes Milhouse will bet only when $Z > \tilde{z}_2$, where $\tilde{z}_2$ solves

$$(2+b)\mathbb{P}[\, M \text{ wins} \mid Z = z] - (1+b)\mathbb{P}[\, M \text{ loses} \mid Z = z] = -1$$

with

$$\mathbb{P}[\, M \text{ wins}] = \frac{\int_0^{\tilde{z}} g_{MB}(x \mid L \text{ folds}) \, dx}{\int_0^1 g_{MB}(x \mid L \text{ folds}) \, dx}.$$

So, for the cases relevant to Bart's decision, Bart has derived the rules he believes Milhouse will use.

Now Bart considers Lisa's situation. He need not calculate Lisa's expected payoff when he folds, since he knows he will lose his dollar. If he thinks she bets with probability $g_L(y)$, the expected value of her game is

$$\mathbb{E}[V_y] = -[1 - g_L(y)] + g_L(y) \{2(1+b)\mathbb{P}[\, M \text{ bets, L wins}] + \tag{4.12}$$
$$(2+b)\mathbb{P}[\, M \text{ folds and L wins}] - (1+b)\mathbb{P}[\, M \text{ bets and L loses}] -$$
$$(1+b)\mathbb{P}[\, M \text{ folds and L loses}]\}.$$

Therefore, to discover her betting rule, Bart must analyze what Lisa thinks about each of the probabilities in (4.12).

Bart believes that Lisa thinks that Milhouse's betting function, when both she and Bart bet, is $g_{LM}(z \mid B \text{ and } L \text{ bet})$. Also, Bart believes that Lisa thinks that Bart's betting function is $g_{LB}(x)$. So Lisa should calculate

$$\mathbb{P}[\, M \text{ bets, L wins}] = \mathbb{P}[\, L \text{ wins} \mid B \text{ and } M \text{ bet}] \mathbb{P}[\, M \text{ bets} \mid B \text{ and } L \text{ bet}]$$
$$= \frac{\int_0^y g_{LB}(x \mid B \text{ bets}) \, dx}{\int_0^1 g_{LB}(x \mid B \text{ bets}) \, dx} \times \frac{\int_0^y g_{LM}(z \mid B \text{ and } L \text{ bet}) \, dz}{\int_0^1 g_{LM}(z \mid B \text{ and } L \text{ bet}) \, dz}(1 - \tilde{z}_3),$$

where $g_{LB}(x \mid B \text{ bets})$ is what Bart thinks Lisa thinks is his betting function, g_{LM} is what Bart thinks Lisa thinks is Milhouse's betting function, and $\tilde{z}_3$ is the threshold that Lisa thinks Milhouse uses when deciding whether or not to bet.

To calculate $\tilde{z}_3$, Bart thinks Lisa reasons as he did in (4.11). She thinks Milhouse will bet when $Z > \tilde{z}_3$, where $\tilde{z}_3$ solves

$$2(1+b)\mathbb{P}[\, M \text{ wins} \mid Z = z] - (1+b)\mathbb{P}[\, M \text{ loses} \mid Z = z] = -1$$

and

$$\mathbb{P}[\, M \text{ wins} \mid Z = z] = \frac{\int_0^{\tilde{z}} g_{LMB}(x) \, dx}{\int_0^1 g_{LMB}(x) \, dx} \times \frac{\int_0^{\tilde{z}} g_{LML}(y \mid B \text{ bets}) \, dx}{\int_0^1 g_{LML}(y \mid B \text{ bets}) \, dy},$$

for $g_{LMB}(x)$ and $g_{LML}(y)$ the betting functions that Bart thinks Lisa thinks Milhouse has for himself and herself, respectively.

Similarly, Bart thinks Lisa finds

$$\mathbb{P}[\text{ M folds, L wins }] = \frac{\int_0^y g_{LB}(x \mid \text{B bets }) \, dx}{\int_0^1 g_{LB}(x \mid \text{B bets }) \, dx} \times \tilde{z}_3$$

$$\mathbb{P}[\text{ M bets, L loses }] = (1 - \tilde{z}_3) - \mathbb{P}[\text{ M bets, L wins }]$$

$$\mathbb{P}[\text{ M folds, L loses }] = \tilde{z}_3 - \mathbb{P}[\text{ M folds, L wins }].$$

Bart has found the quantities Lisa uses to solve (4.12), and can solve to get the value $\tilde{y}$ that is her threshold for betting.

Now Bart has what he needs to determine the strategy that maximizes his expected value. He should bet with $X = x$ when the expected value is greater than -1, the amount that he loses by folding; i.e., he bets when

$$-1 < 2(1+b)\mathbb{P}[\text{ B wins and L, M bet }] + (2+b)\mathbb{P}[\text{ B wins and only L bets }] +$$
$$(2+b)\mathbb{P}[\text{ B wins and only M bets }] - (1+b)\mathbb{P}[\text{ B loses and L, M bet }] -$$
$$(1+b)\mathbb{P}[\text{ B loses and only L bets }] - (1+b)\mathbb{P}[\text{ B loses and only M bets }].$$

He knows that Lisa bets when $Y > \tilde{y}$, and that Milhouse bets when either $Z > \tilde{z}_1$ and Lisa bets, or when $Z > \tilde{z}_2$ and Lisa folds.

From this information, given $X = x$,

$$\mathbb{P}[\text{ B wins and L, M bet }] = \mathbb{P}[\text{ B wins } \mid \text{ L, M bet }] \mathbb{P}[\text{ L, M bet }]$$
$$= \mathbb{P}[\text{ B wins } \mid \text{ L, M bet }] (1 - \tilde{y})(1 - \tilde{z}_1)$$
$$= \frac{[x - \max\{\tilde{y}, \tilde{z}_1\}]^+}{(1 - \max\{\tilde{y}, \tilde{z}_1\})} (1 - \tilde{y})(1 - \tilde{z}_1).$$

Similarly,

$$\mathbb{P}[\text{ B wins and only L bets }] = \frac{[x - \tilde{y}]^+}{1 - \tilde{y}} (1 - \tilde{y}) \tilde{z}_1$$

$$\mathbb{P}[\text{ B wins and only M bets }] = \frac{[x - \tilde{z}_2]^+}{1 - \tilde{z}_2} \tilde{y}(1 - \tilde{z}_2).$$

Each of these terms can be computed, and the remaining probability terms can be found from the fact that $\mathbb{P}[A \text{ and } B^c] = \mathbb{P}[A] - \mathbb{P}[A \text{ and } B]$, where the superscript denotes complementation.

The previous ARA analysis may seem more complex than it is. Most of the effort is simply a matter of bookkeeping. One must keep track of what Bart thinks each player believes about each of the other players.

Exercises

4.1. A Defender will invest $x > 0$ amount of resources at a site to protect it against attack. The Attacker observes her investment, and chooses whether or not to attack. The site has value v to whichever player ultimately controls it. The cost of an attack is $0 < c_A < v$. The unit cost of the defensive resources is $0 < c_D < v$, which may be deployed by the Defender in fractions of a unit. If the site is attacked, there is probability $x/(x+1)$ that the Defender will prevail. If the defender and the attacker are both risk neutral, and assuming common knowledge, what is the standard game-theoretic solution?

4.2. Consider a Defender who must allocate limited resources between two sites that are possible terrorist targets. The terrorists have only enough resources to attack one of these two sites, and they will decide their target after observing the defender's allocation. Assume that (i) an attack on a site without defensive resources is always successful, and (ii) the probability of an attack being successful decreases proportionally with the percentage of defensive resources allocated to a site. In particular, an attack on a site that receives all the defender's resources has 0-probability of succeeding. Suppose Defender and Attacker are risk neutral, and that this problem is a zero-sum game. Find the defender's optimal resource allocation as a function of the sites' values.

4.3. Solve Exercise 4.2 when the effectiveness of defenses in reducing the attack success probability may depend on the site where resources are allocated. Specifically, suppose that each site may need a different amount of protection to produce a fixed probability of success in its vulnerability to attacks. Interpret the solution for when both sites have the same value.

4.4. Suppose the defender in Exercise 4.2 gets intelligence on the probability of an attack on site #1 versus site #2, and that based on this information she models the Attacker as a 0-level thinker. What resource allocation between sites are recommended by such an approach? Compare the results with those in Exercise 4.3 and evaluate the appropriateness of this intelligence in predicting the attacker's choice.

Chapter 5
Variations on Sequential Defend-Attack Games

Previously, we described the adversarial risk analysis (ARA) approach to simultaneous games and sequential games. We now show how ARA applies when there are complicating factors, such as multiple defenders, multiple attackers, and multiple targets to protect. We also treat situations in which there are varying degrees of cooperation among attackers or among defenders, focusing on the sequential Defend-Attack game, although the ideas apply to many other models presented in this book. These games were introduced in Chapter 4, and have been studied from a classical game-theoretic perspective by Bier and Azaiez (2009) and Brown et al. (2006).

5.1 The Sequential Defend-Attack Model

Recall that in the basic sequential Defend-Attack game, the Defender first chooses a defense and, then, having observed it, the Attacker chooses an attack. To simplify the discussion, assume that the Defender has a discrete set of possible defenses $\mathcal{D} = \{d_1, d_2, ..., d_m\}$ and she must choose one of these. Similarly, the Attacker has a set of possible attacks $\mathcal{A} = \{a_1, a_2, ..., a_n\}$ from which he must choose one. (The choice sets may include the option of doing nothing, or combining several defenses or several attacks.) For further simplicity, assume that the only uncertainty is a binary outcome $S \in \{0, 1\}$, which represents whether or not the attack was successful. Finally, for both adversaries, their utilities depend only upon S and the action they chose. (In some cases, the utility might also depend upon the action chosen by the opponent, as in asymmetric warfare, where the weaker party seeks to leverage disproportionate investment by the stronger.)

Figure 5.1 depicts the problem graphically. The top image is a MAID, showing interlocked influence diagrams for both participants with a shared uncertainty node and a linking arrow. The influence diagram explicitly shows that the uncertainty associated with the success S of an attack has probabilistic dependence on the actions of both the Attacker and the Defender. Arcs that point to a utility node represent functional dependence (Shachter, 1986). Thus, the utility functions over the consequences for the Defender and the Attacker are, respectively, $u_D(d,S)$ and $u_A(a,S)$. The arc in the influence diagram from the Defender's decision node to the Attacker's shows that the Defender's choice is observed by the Attacker. The lower image in Fig. 5.1 is a game tree for the Defend-Attack problem (with only two actions per adversary: $m=n=2$). It more clearly depicts the turn-taking in the decision problem. Note that there are two utility values, for the Attacker and for the Defender, at the terminal nodes of the tree.

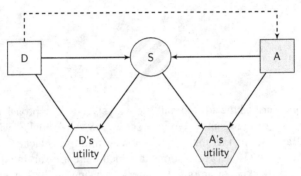

(a) Influence diagram for the sequential Defend-Attack game.

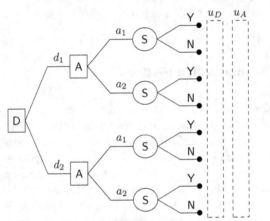

(b) Game tree for the sequential Defend-Attack game ($m=n=2$).

Fig. 5.1 Two representations of the sequential Defend-Attack game.

Nearly all previous work on the sequential Defend-Attack problem uses standard game theory, and it is an example of the well-studied class of Stackelberg games (cf. Aliprantis and Chakrabarti, 2010). The solution requires a probability assessment over S, conditional on (d,a), the choices made by the Defender and the Attacker. The Defender and the Attacker may have different assessments for the chance of success, and these are represented by $p_D(S=1 \mid d,a)$ and $p_A(S=1 \mid d,a)$, respectively. The solution does not require the Attacker to know the Defender's probabilities and utilities, since he observes her actions, but the Defender needs to know the Attacker's probabilities and utilities.

To solve the problem, one needs the expected utilities of the players at node Ⓢ of the game tree in Figure 5.1. The expected utility that the Attacker obtains when the decisions are $(d,a) \in \mathcal{D} \times \mathcal{A}$ is

$$\psi_A(d,a) = p_A(S=0 \mid d,a)\, u_A(a, S=0) + p_A(S=1 \mid d,a)\, u_A(a, S=1). \quad (5.1)$$

One computes $\psi_D(d,a)$ similarly for the Defender. The Attacker's best attack against the defense d is

$$a^*(d) = \mathrm{argmax}_{a \in \mathcal{A}}\, \psi_A(d,a), \ \forall\, d \in \mathcal{D}. \quad (5.2)$$

Under the assumption that the Defender knows how the Attacker will solve his problem, then her best defense is

$$d^* = \mathrm{argmax}_{d \in \mathcal{D}}\, \psi_D(d, a^*(d)).$$

The solution $(d^*, a^*(d^*))$ is a Nash equilibrium, and, more specifically, a subgame perfect equilibrium (Menache and Ozdaglar, 2011).

If the Defender accurately knows the Attacker's true p_A and u_A, then the Defender has the advantage: she is the first to move and, *ceteris paribus*, has larger expected utility than in the simultaneous Defend-Attack game described in Chapter 2. Bier and Azaiez (2009) discusses this in detail, pointing out that sometimes disclosing information about defenses against terrorism is a deterrent. However, the assumption that the Defender precisely knows the Attacker's preferences and probability assessments is usually unrealistic.

ARA allows the analyst to avoid that problematic assumption, at the usual price of having to rely upon subjective probability assessments. When the Defender does not actually know (p_A, u_A), then one can view the Defender's situation as a standard decision analysis problem: the Defender's influence diagram in Figure 5.2 removes the hexagonal utility node containing the Attacker's information and replaces the decision rectangle $\boxed{A}$ with the circle Ⓐ to indicate that the Attacker's choice is a random variable.

The influence diagram shows that the Defender knows her own utilities and probabilities, but not those of the Attacker. She must assess her beliefs about $p_D(A \mid d)$, which is the probability that the Attacker will choose attack a after observing that the Defender has chosen defense d. This assessment requires the Defender to analyze the problem from the Attacker's perspective, and ARA allows this to be done in several possible ways.

For example, the Defender may believe the Attacker is non-strategic, in which case she places a subjective distribution over his actions. Or she may think he maximizes his expected utility conditional on her observed defense choice d, in which case she needs to use all available information to assess his utilities and probabili-

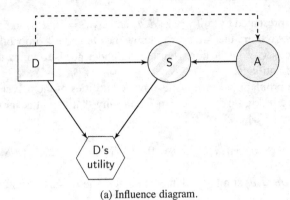

(a) Influence diagram.

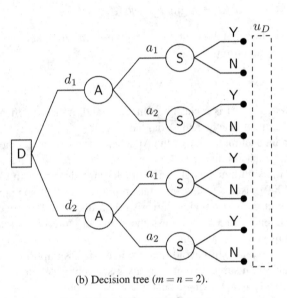

(b) Decision tree ($m = n = 2$).

Fig. 5.2 Two representations of the Defender's decision problem.

ties in order to solve the problem represented in Fig. 5.2. And if she thinks he is a prospect maximizer, then she would proceed similarly, but reason about value and weighting functions instead of utilities.

Suppose the Defender thinks the Attacker maximizes his expected utility. To find $p_D(A \mid d)$, the Defender first estimates the Attacker's utility function and his probabilities for success S, conditional on (d, a), and consequently computes the required probability. However, instead of using point estimates for p_A and u_A to find the Attacker's optimal decision $a^*(d)$ as in (5.2), the Defender's uncertainty about the Attacker's decision should derive from her uncertainty about the Attacker's (p_A, u_A), which she describes through a distribution F. This, in turn, induces a distribution on the Attacker's expected utility $\psi_A(a, d)$ defined in (5.1). Thus, if the Attacker is an expected utility maximizer, the Defender's distribution about the Attacker's decision, given her defense choice d, is

$$p_D(A = a^* \mid d) = \mathbb{P}_F[a^* = \text{argmax}_{a \in \mathcal{A}} \Psi_A(d,a)],$$

where, for $(P_A, U_A) \sim F$,

$$\Psi_A(d,a) = P_A(S = 0 \mid d,a) \, U_A(a, S = 0) + P_A(S = 1 \mid d,a) \, U_A(a, S = 1).$$

For each d, the Defender may use Monte Carlo simulation to approximate $p_D(A \mid d)$ by drawing n samples $\left\{ (p_A^i, u_A^i) \right\}_{i=1}^n$ from F, which produce $\{\psi_A^i\}_{i=1}^n \sim \Psi_A$. She then approximates $p_D(A = a \mid d)$ by

$$\hat{p}_D(A = a \mid d) = \frac{\#\{a = \text{argmax}_{x \in \mathcal{A}} \, \psi_A^i(d,x)\}}{n}, \quad \forall a \in \mathcal{A}.$$

Once the Defender has completed her assessments, she can solve her problem. Her expected utilities at node ⑤ in Figure 5.2 for each $(d,a) \in \mathcal{D} \times \mathcal{A}$ are

$$\psi_D(d,a) = p_D(S = 0 \mid d,a) \, u_D(d, S = 0) + p_D(S = 1 \mid d,a) \, u_D(d, S = 1).$$

Then, her estimated expected utilities at node Ⓐ for each $d \in \mathcal{D}$ are

$$\hat{\psi}_D(d) = \sum_{i=1}^k \psi_D(d, a_i) \, \hat{p}_D(A = a_i \mid d).$$

Finally, her optimal decision is $d^* = \text{argmax}_{d \in \mathcal{D}} \, \hat{\psi}_D(d)$.

Note that, in contrast with classic game theory, the solution d^* for the sequential game need not correspond to a Nash equilibrium. Assume there is a third party who knows both the Defender's and the Attacker's true probabilities and utilities, as well as the Defender's belief F about the Attacker's utilities and probabilities. That third party could then predict the game, identifying the decisions made by each player. However, this omniscient prediction is unlikely to coincide with the Nash equilibrium derived from the actual (p_D, u_D) and (p_A, u_A), since the players lack full and common knowledge.

ARA requires assessment of F. Regarding the random probabilities $P_A(s \mid d,a)$, she could base them on her own assessments $p_D(s \mid d,a)$, but with additional noise to reflect her uncertainty. When S is binary (success or failure), she could use a beta distribution chosen so that $P_A(s \mid d,a)$ concentrates on her known $p_D(s \mid d,a)$. If S is discrete, her draws could have the Dirichlet distribution centered at $p_D(s \mid d,a)$ with variance determined by her certitude. And if S is continuous, then she would use the analogous Dirichlet process (Ferguson, 1973).

For the random utility $U_A(d,s)$, the Defender must generally study whatever information she has about the aims of the attackers, as in Keeney (2007), Keeney and von Winterfeldt (2010), and Keeney and von Winterfeldt (2011). Her research could lead to a model based on a weighted measurable value function, as in Dyer and Sarin (1979). To take account of risk attitudes, she might appeal to the relative risk-aversion concept (cf. Dyer and Sarin, 1982), and assume the attackers are risk prone. Typically, there would be uncertainty about the weights of the value function and the risk proneness coefficient, which would be modeled through random variables, inducing a random utility model.

Wang and Bier (2013) provides an alternative approach for assessing adversary utilities. It makes ordinal judgments about the preferences and then applies the probabilistic inversion method to create approximate utilities (cf. Kraan and Bedford, 2005).

5.2 Multiple Attackers

Aside from brief comments in the discussion of auctions and *La Relance*, all previous examples concerned cases in which there was a single opponent. But in the sequential Defend-Attack game, and more generally in any adversarial situation, it is entirely plausible to face more than one Attacker, and these opponents may have different sets of resources, different goals, and different degrees of cooperation. For example, governments must simultaneously defend against state-sponsored terrorism, franchise terrorists, and solitary actors; similarly, police must defend against vandals, gangs, and organized crime.

First, consider the situation in which attacks are not coordinated. One example is shown by the MAID in Figure 5.3, which assumes there are two Attackers. The lack of coordination is indicated by the absence of an arc between A_1 and A_2. Note, however, that the outcomes of the actions of both attackers affect the results; this is shown, e.g., in the fact that S_1 depends on D, A_1, and A_2.

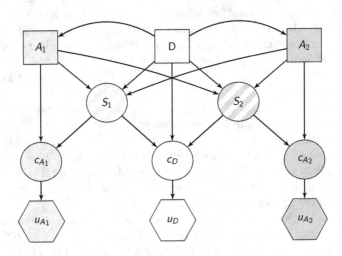

Fig. 5.3 Multiagent influence diagram for a bithreat problem.

In general, the Defender must deploy defensive resources $d \in \mathcal{D}$ to face K uncoordinated attackers $A_1, \ldots, A_K$. These observe her decision, and then choose attacks a_k from among, respectively, $\mathcal{A}_k$ for $k = 1, \ldots, K$. The interaction between the Defender D and each Attacker A_k is through their respective decisions d and a_k, and leads to the random result $S_k \in \mathcal{S}_k$, which depends on all decisions that are made (i.e., the probability of a successful attack by A_k with choice a_k is affected by A_j's choice a_j, and vice versa). The Defender obtains multi-attribute consequences c_D; e.g., she may count deaths, injuries, and financial damage separately, but later combine these in her utility function. Her utility $u_D(d, \mathbf{s})$ depends upon both her decision d and the success of the attacks $\mathbf{s} = (s_1, \ldots, s_K)$, which jointly determine the consequences. Similarly, each attacker obtains multi-attribute consequences c_{A_k}, which depend on both his choice a_k and the result s_k. These consequences are combined in his utility function $u_{A_k}(a_k, s_k)$, which depends upon the actions of other Attackers only through the success or failure of his own attack.

The Defender wants to find her optimal defense d^*. If the outcomes S_k of different attacks are independent, conditional on her choice d and the chosen attacks $\mathbf{a} = (a_1, \ldots, a_K)$, she can assess her probabilities $p_D(s_k \mid d, \mathbf{a})$ for the outcome of attack a_k given the choices made by all players. Note that, in this formulation, her probability for the success of attack a_k does not depend on whether the other attacks were successful, but only upon which choices the other Attackers made. This is a reasonable approximation when the Defender is highly resourced and the Attackers do not coordinate. For example, the outcome of one burglary attempt is probably not affected by whether or not other burglars are successful, but it may be affected by the fact that other people choose to burgle; i.e., if a neighborhood sees a rash of attempted robberies, then police increase their surveillance, lowering each burglar's chance of success. But if the Defender is not sufficiently well resourced, this framing of the problem is less plausible—an understaffed police department means that a successful attack diverts resources, increasing the chance that other attacks will be successful. Similarly, if Attackers coordinate, so that multiple burglaries occur simultaneously, this increases the chance of successful burglary for all.

Given her probabilities, the Defender calculates her expected utility for the attacks, integrating out her uncertainty about the outcome of each attack:

$$\psi_D(d \mid \mathbf{a}) = \int_{\mathcal{S}} u_D(d, \mathbf{s}) \prod_{k=1}^{K} p_D(s_k \mid d, \mathbf{a}) \, d\mathbf{s}, \tag{5.3}$$

where $\mathcal{S} = \mathcal{S}_1 \times \cdots \times \mathcal{S}_K$. Now suppose that the Defender is able to find the models $p_D(a_k \mid d)$, $k = 1, \ldots, K$, which express her beliefs about which attack will be chosen by the kth attacker, after he has observed her defense choice d. Under the assumption of uncoordinated attacks, so that $a_1, \ldots, a_K$ are independent given d, the Defender can then compute

$$\psi_D(d) = \int_{\mathcal{A}} \psi_D(d \mid \mathbf{a}) \prod_{k=1}^{K} p_D(a_k \mid d) \, d\mathbf{a},$$

where $\mathcal{A} = \mathcal{A}_1 \times \cdots \times \mathcal{A}_k$, and then determine

$$d^* = \text{argmax}_{d \in \mathcal{D}} \, \psi_D(d)$$

to find her optimal defense choice.

In solving this problem, the Defender had to assess the utilities $u_D(d, s)$, the distributions $p_D(s_k \mid d, \boldsymbol{a})$ and the distributions $p_D(a_k \mid d)$, $k = 1, \ldots, K$. To obtain these, she proceeds as in Section 5.1. The Defender must model the thinking of each attacker, and she should solve their corresponding problems separately, since the Attackers do not coordinate.

Sometimes it is reasonable to suppose that uncoordinated attacks have independent effects. For example, this would occur if one Attacker is a murderer and the other Attacker is a burglar, where the police department had separate divisions to handle those crimes. The outcome for the attempted murder would not affect the outcome for the attempted burglary, and so $\prod_{k=1}^{K} p_D(s_k \mid d, \boldsymbol{a})$ in (5.3) is replaced by $\prod_{k=1}^{K} p_D(s_k \mid d, a_k)$. This situation is shown by the MAID in Fig. 5.4, which does not have arcs between each Attacker's box to the other Attacker's chance node. Here, the Defender's action is to decide how much money to allocate to the Homicide Unit and to the Burglary Unit.

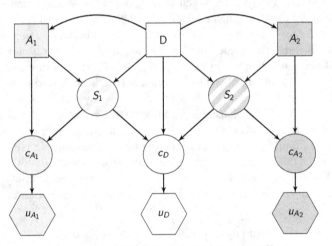

Fig. 5.4 The bithreat problem when multiple attacks have independent effects.

In other cases, attacks can have cascade effects, by inspiring others to take action. This pattern is apparent in franchise terrorism, suicide bombings, and school shootings, and corresponds to very weak coordination. A stronger form of coordination occurs if the Attackers agree to order their decisions, so that A_1 chooses first, A_2 chooses second, and so forth. In that case one replaces the product $\prod_{k=1}^{K} p_D(s_k \mid d, \boldsymbol{a})$ in (5.3) with

$$p_D(s_1 \mid d,a_1)p_D(s_2 \mid d,a_1,a_2)\cdots p_D(s_K \mid d,a_1,\ldots,a_K).$$

This assumption of turn-taking is implausible in the context of terrorism or crime, but it exactly describes multiplayer *La Relance* in Section 4.4.2, where the coordination among the Attackers is the order of play: Lisa bets after Bart, and Milhouse bets after Lisa. A three-person turn-taking game, where the Defender moves, then the Attackers choose simultaneously but Attacker 1 moves first, and the success of the second attack depends upon the outcome of the first, is shown by the MAID in Fig. 5.5. The arc from $\textcircled{S_2}$ to $\textcircled{S_1}$ indicates the dependence upon the past.

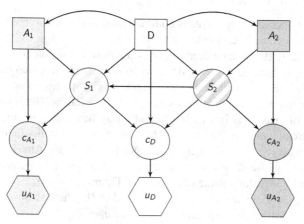

Fig. 5.5 A bithreat problem with cascading effect.

For multiplayer *La Relance*, there is just one random outcome S which depends upon the decisions of all the players. Also, the analysis is affected only by the sequence of decisions to bet or to fold, and not the ultimate outcome. But in terrorism, it is quite possible that the decision to attack depends not only upon whether other agents have launched previous attacks, but also upon the success of those attacks. For such cases, the product in (5.3) should be

$$p_D(s_1 \mid d,a_1)p_D(s_2 \mid d,a_1,a_2,s_1)\cdots p_D(s_K \mid d,a_1,\ldots,a_K,s_1,\ldots,s_{K-1}).$$

In general, the treatment of partial coordination among multiple Attackers is difficult. If there is no strict turn-taking, then one needs to extend the action sets to include choices about timing. And if multiple attacks can weaken defenses, so that the damage done by a specific attack depends upon how many other attacks were successful, then a more complicated model is required. This will be lightly addressed in the case study in Chapter 6, in which pickpocketing reduces the sense of security in a railway system, leading to decreased ridership.

When multiple opponents coordinate their attacks, the ARA for the sequential Defend-Attack game must take account of the kind of cooperation that exists. The previous discussion treated examples of weak coordination, through attack cascades inspired by previous attacks and turn-taking. But often attacks are strongly coordinated and explicitly strategic. Important examples include:

- The Spanish government needed to defend its people against the joint operations of the Basque terrorist organization ETA, the 'Ndrangheta and the Colombian narcotics mafia.
- Iran is the subject of coordinated economic sanctions imposed by many other (mostly Western) nations.
- Exxon must compete in a world market in which critical price controls are imposed by the OPEC.

Combating cooperative opponents is common, and arises in terrorism, international relations, business, and many other circumstances.

When there is strong coordination, one can view the problem as one in which there is a single Attacker. The challenge is to determine the group utility function that represents the shared interest of the Attackers. Group utilities combine the individual utility functions of each Attacker. Keeney and Raiffa (1993) and Ríos Insua et al. (2008) discuss ways in which different utility functions can combine. For example, if one can monetize the costs and benefits, the group utility function might be the maximum loss or minimum gain experienced by any of the Attackers. In the context of bargaining, the algorithms often correspond to group value functions which aggregate multiple distinct value functions (cf. Thomson, 1994).

As before, the Defender chooses an action $d \in \mathcal{D}$, which is observed by the K Attackers. The Attackers respond with choices $a_k \in \mathcal{A}_k$, $k = 1, \ldots, K$, leading to random outcomes $S_k \in \mathcal{S}_k$ with multi-attribute consequences c_D for the Defender and c_k for each Attacker. Since c_k potentially depends upon the actions $a_1, \ldots, a_K$ and the outcomes $s_1, \ldots, s_K$, it is concise to let $u_G(c_1, \ldots, c_k)$ denote the group utility function for the Attackers.

The Defender seeks her optimal defense strategy d^*. To find it she must assess her probabilities $p_D(s \mid d, a)$ and calculate her expected utility as

$$\psi_D(d \mid a) = \int_{\mathcal{S}} u_D(d,s) \, p_D(s \mid d, a) \, ds.$$

If the Defender is also able to assess her beliefs $p_D(a \mid d)$ about the probability of each set of Attacker actions conditional on her defense choice, she can then find

$$\psi_D(d) = \int_{\mathcal{A}} \psi_D(d \mid a) \, p_D(a \mid d) \, da,$$

and solve $d^* = \text{argmax}_{d \in \mathcal{D}} \, \psi_D(d)$ to obtain her best defense.

The key problem in the Defender's ARA is the assessment of $p_D(a \mid d)$. If the attackers are expected group utility maximizers (cf. Keeney and Raiffa, 1993) and if she believes that they have a consensus on group probabilities $p_G(s \mid d, a)$ for the outcomes given the choices, then she would solve

$$a^*(d) = \text{argmax}_{\boldsymbol{a} \in \mathcal{A}} \int_{\mathcal{S}} u_G(c_1, \ldots, c_K) \, p_G(\boldsymbol{s} \mid d, \boldsymbol{a}) \, d\boldsymbol{s},$$

where the c_k depend on all of the Attacker actions and on the outcomes $\boldsymbol{s}$.

The Defender does not know u_G and p_G, she expresses her uncertainty through random utilities and probabilities (U_G, P_G). She can propagate this uncertainty to obtain the random optimal attack, given her defense d:

$$A^*(d) = \text{argmax}_{\boldsymbol{a} \in \mathcal{A}} \int_{\mathcal{S}} U_G((c_1, \ldots, c_K) \, P_G(\boldsymbol{s} \mid d, \boldsymbol{a}) \, d\boldsymbol{s}.$$

She now obtains $p_D(\boldsymbol{a} \mid d) = \mathbb{P}[A^*(d) = \boldsymbol{a}]$, which can be approximated by simulation, as in Section 5.1.

For the distribution on P_G, the Defender may use Dirichlet distributions or processes, as appropriate, and these are often centered on the Defender's own beliefs about the probabilities of the outcomes. For the distribution on U_G she could take it to be the maximum loss or minimum gain. Esteban and Ríos Insua (2014b) suggest letting $u_k(a_k, s_k)$, $k = 1, \ldots, K$ be point estimates for the utilities obtained by each of the attackers, and letting $(t_1, \ldots, t_K)$ be tolerances. They then define $u_G = [(u_1 - t_1)^q + \cdots + (u_K - t_K)^q)]^{1/q}$, as the group utility function, where randomness enters through a distribution over q.

5.3 Multiple Defenders

Besides multiple attackers, it is also common to have multiple defenders. Examples include:

- Different banks, which share information in order to create credit scores for customers that reduce default rates and fraud.
- Airline companies, which each operate baggage screening systems to prevent the introduction of bombs onto planes, but who must generally trust the security systems of each other when transferring luggage between carriers (Kunreuther and Heal, 2003).
- The multinational military effort in Afghanistan, where the United States, Great Britain, Canada, and others seek to protect the government from overthrow by the Taliban.

Note that each defender may each be protecting its own targets (as in the case of several countries, each defending its people against al Qaeda), or they may defend a common target (as when several companies jointly invest in computer security to protect a database that all use). These defenders may act independently, or with weak or strong coordination.

First consider the case of L uncoordinated Defenders D_ℓ, $\ell = 1, \ldots, L$, that face an Attacker A. Examples include homeowners who decide whether or not to invest in burglar alarms, irrespective of what decision their neighbors make, or companies that protect their intranets from cyberattacks, irrespective of what other companies do, and countries that decide to protect their ports from smuggled nuclear devices, regardless of what security procedures other countries undertake.

This ARA supports defender D_1, who needs to deploy defensive resources $d_1 \in \mathcal{D}_1$ to face an attacker A, in the company of the other $L-1$ defenders, who make their corresponding decisions $d_\ell, \ell = 2, \ldots, L$. The attacker observes all defensive choices, and then selects his attack $a \in \mathcal{A}$. The interaction between D_ℓ and A through their respective decisions d_ℓ and a, leads to a random result $S_\ell \in \mathcal{S}_\ell$. The ℓth Defender faces multi-attribute consequences c_ℓ, which depend on her defense d_ℓ and the results s_ℓ. She then gets her utility $u_\ell(d_i, s_i)$. The attacker will get his multi-attribute consequences c_A, which depend on his attack effort a and the results $\boldsymbol{s} = (s_1, \ldots, s_L)$. His utility is $u_A(a, \boldsymbol{s})$.

In this formulation of the game, the utility received by one Defender does not depend on the outcomes for the other Defenders. That assumption is reasonable in the context of, say, purchase of a burglar alarm, but less plausible in terms of screening baggage, since any successful bombing of an airplane impacts the entire industry. However, it is straightforward to extend the following ARA to the case in which a Defender's utility is affected by the outcomes for other Defenders—one simply needs to condition on those outcomes, develop personal probabilities applicable to that model, and then integrate over all of the outcomes.

The first Defender seeks her optimal defense d_1^*. For this, she computes her expected utility, conditional on what she believes the other defenders and the attacker will do:

$$\psi_1(d_1 \mid d_2, \ldots, d_L, a) = \int u_1(d, s_1) p_1(s_1 \mid \boldsymbol{d}, a) ds_1,$$

where $\boldsymbol{d} = (d_1, \ldots, d_L)$. Then, based on her forecast of what the other defenders will do, and what the attacker will do given the deployed defenses, she computes

$$\psi_1(d_1) = \int_{\mathcal{A}} \int_{\mathcal{D}_{-1}} \psi_1(d_1 \mid \boldsymbol{d}_{-1}, a) p_1(a \mid \boldsymbol{d}) p_1(\boldsymbol{d}_{-1}) d\boldsymbol{d}_{-1} da,$$

where $\boldsymbol{d}_{-1} = (d_2, \ldots, d_L)$ and $\mathcal{D}_{-1} = \mathcal{D}_2 \times \cdots \times \mathcal{D}_L$. She now solves for

$$d_1^* = \text{argmax}_{d \in \mathcal{D}_1} \; \psi_1(d),$$

which is her best defense.

To perform this analysis, she needs to assess $p_1(a \mid \boldsymbol{d})$ and $p_1(\boldsymbol{d}_{-1})$. For the first distribution, she knows that the (rational) Attacker's choice should be

$$a^*(\boldsymbol{d}) = \text{argmax}_{a \in \mathcal{A}} \int_{\mathcal{S}} u_A(a, \boldsymbol{s}) p_A(\boldsymbol{s} \mid \boldsymbol{d}, a) d\boldsymbol{s},$$

where $\mathcal{S} = \mathcal{S}_1 \times \cdots \times \mathcal{S}_L$. But the Defender does not know the Attacker's u_A nor p_A. As in the discussion of multiple Attackers, she uses random utilities and probabilities (U_A, P_A) to express her uncertainty, solving

$$A^*(\boldsymbol{d}) = \operatorname{argmax}_{a \in \mathcal{A}} \int_{\mathcal{S}} U_A(a,\boldsymbol{s}) P_A(\boldsymbol{s} \mid \boldsymbol{d},a) \, d\boldsymbol{s} \qquad (5.4)$$

to find the distribution of the optimal attack, which to her is a random variable. Her distribution for that random variable can be determined by repeated sampling of (U_A, P_A) and solution of (5.4). Often, a reasonable structural assumption is that

$$p_A(\boldsymbol{s} \mid \boldsymbol{d},a) = \prod_{\ell=1}^{L} p_A(s_\ell \mid d_\ell, a).$$

Defender D_1 also needs to assess $p_1(\boldsymbol{d}_{-1})$, which describes her belief about how the other Defenders will act. Because of the simultaneous play and non-coordination among the Defenders, she may reasonably assume that their choices are independent. Thus $p_1(\boldsymbol{d}_{-1}) = \prod_{\ell=2}^{L} p_1(d_\ell)$. To determine, say, $p_1(d_2)$, she would step into the shoes of D_2, who must solve

$$\psi_2(d_2 \mid \boldsymbol{d}_{-2}, a) = \int_{\mathcal{S}_2} u_2(d_2, s_2) p_2(s_2 \mid \boldsymbol{d}, a) \, ds_2,$$

where $\boldsymbol{d}_{-2} = (d_1, d_3, \ldots, d_L)$. This enables D_2 to find

$$\psi_2(d_2) = \int_{\mathcal{A}} \int_{\mathcal{D}_{-2}} \psi_2(d_2 \mid \boldsymbol{d}_{-2}, a) p_2(a \mid \boldsymbol{d}) p_2(\boldsymbol{d}_{-2}) \, d\boldsymbol{d}_{-2} \, da,$$

where $\mathcal{D}_{-2} = \mathcal{D}_1 \times \mathcal{D}_3 \times \cdots \mathcal{D}_L$. Then the optimal play for D_2 is

$$d_2^* = \operatorname{argmax}_{d \in \mathcal{D}_2} \psi_2(d).$$

But, as usual, D_1 does not know $u_2(d_2, s_2)$, $p_2(a \mid \boldsymbol{d})$, nor $p_2(\boldsymbol{d}_{-2})$. So D_1 places a subjective distribution over $U_2(d_2, s_2)$, $P_2(a \mid \boldsymbol{d})$, and $P_2(\boldsymbol{d}_{-2})$, analogously with previous inference, and uses repeated sampling to find the distribution of

$$D_2^* = \operatorname{argmax}_{d \in \mathcal{D}_2} \int_{\mathcal{A}} \int_{\mathcal{D}_{-2}} U_2(d_2, s_2) P_2(s_2 \mid \boldsymbol{d}, a) P_2(a \mid \boldsymbol{d}) P_2(\boldsymbol{d}_{-2}) \, d\boldsymbol{d}_{-2} \, da.$$

The same reasoning would apply for each of the other Defenders.

Note that in applying this logic one could start a recursion based on what the other Defenders think is the modeling being performed by a given Defender. This enables use of the kind of level-k analyses discussed in Section 3.6, in the context of auctions among three bidders.

When the Defenders coordinate, one must distinguish complete cooperation from partial cooperation. Examples of complete cooperation include soldiers in a combat squadron, or a neighborhood association that requires all households to contribute a fixed amount to provide security. In these situations, there is centralized decision-making and so one may view the problem as a two-person sequential Defend-Attack game, as analyzed in Section 5.1.

Partial cooperation is more complicated. A prominent example is the international military alliance between the United States, the United Kingdom, Australia, and Poland which led to the invasion of Iraq in 2003. Each nation had different interests and made different contributions, but there was sufficient common ground and negotiated structure that mutual choices were made. A looser level of coordination occurs in, say, a Neighborhood Watch program, for which different individuals volunteer different amounts of time and financial support, and coverage may vary widely, depending upon personal circumstances.

In the ideal cooperative case, each Defender elicits her probabilities about the Attacker's choice, conditional on all sets of defensive actions, and also her probabilities for the outcomes given the attack and the defenses. If each Defender finds that her expected utility is maximized by the same set of joint decisions, then the problem is solved. But such agreement is rare.

An alternative, when there is disagreement, is to compromise. The Defenders could accept any set of decisions for which each Defender receives an increase in her expected utility. If there is more than one set of defense choices that has that property, then the Defenders negotiate, and perhaps decide to use the set that maximizes the minimum gain in expected utility, or which maximizes the average gain in expected utility.

Regrettably, it may often happen that there is no set of decisions that improves the expected utility of all Defenders. In that case hard negotiation is required, and Defenders who anticipate large gains in expected utility must find ways to compensate those who expect a loss. The chosen solution depends sensitively upon the resources and relationships between the Defenders.

There are other options. If none of the Defenders feels confident in her elicited probabilities for the Attacker's choice and/or her elicited probabilities for the outcomes, conditional on all sets of defense choices, then they might regard their probabilities as a draw from a common distribution. By pooling their beliefs, the Defenders could broker agreement on a common set of probabilities, enabling a unified solution as if there were a single Defender. This approach entails combination of subjective beliefs, which is notoriously problematic (Seidenfeld, Kadane and Schervish, 1989) but also often necessary (Das, Yang and Banks, 2013).

If the Defenders can agree on a common $p(s \mid d, a)$, then a different option follows the spirit of Esteban and Ríos Insua (2014b). First, each Defender performs her own analysis, obtaining her maximum expected utility Ψ_ℓ^*, $\ell = 1, \ldots, L$. Next, if the Defenders jointly implement $(d_1, \ldots, d_L)$ and the attacker selects action a, then the random result will be $(S_1, \ldots, S_L)$, and each Defender can calculate her expected utility $\mathbb{E}[u_\ell(d_\ell, S_\ell)]$. Define, then, the cooperation penalty for defender D_ℓ as

$$g_\ell(d) = \max\{\mathbb{E}(u_\ell(d_\ell, S_\ell) - \Psi_i^*), 0\},$$

which may achieve zero for some players when the choice d benefits them, although their individual analyses assigned low probability to the group making that selection. Finally, the Defenders solve

$$d^* = \text{argmax}_{d \in \mathcal{D}} \int_{\mathcal{S}} \left(\sum g_\ell(s_\ell)^q \right)^{1/q} p(s \mid d, a) \, ds$$

which, in this framework, is the optimal group defense decision.

5.4 Multiple Targets

Consider sequential Defend-Attack games in which the Defender must protect multiple targets from a threat. Examples of Defenders in such games include:

- A mayor, who must allocate police resources across multiple precincts, to control several kinds of criminal activity.
- A CEO, who must develop a budget that funds different departments within the organization, where each department (target) faces competition from a competitor (threats).
- A government, which assigns security personnel to embassies in other countries, where its interests may be threatened by bombs, mobs, or espionage.

The game can be seen as an application of portfolio theory in which an opponent observes the Defender's investments and seeks to minimize her return.

Suppose there are I targets and J kinds of resources. The Defender may deploy an amount d_{ij} of the jth resource to protect the ith target, so the entire decision is represented by the matrix $D = \{d_{ij}\}$. These allocations must satisfy two standard constraints:

$$d_{ij} \geq 0, \quad \sum_{i=1}^{I} d_{ij} \leq T_j.$$

The first ensures that negative investment is impossible, and the second implies that there is a ceiling, T_j, on the amount of the jth resource that is available. In general, there are additional constraints, such as the requirement that a bomb detecting dog must always be accompanied by a security officer, or a directive that some targets receive a minimum level of protection. The feasible choice set for the Defender is denoted by $\mathcal{D}$.

In this sequential Defend-Attack game, the Attacker has K resources that may be used for attack. He observes the initial set of investments D and decides to allocate to the ith target an amount a_{ik} of the kth attack resources. Similarly to the Defender, his full decision is represented by a matrix A such that entry $a_{ik} \geq 0$ and $\sum_i a_{ik} \leq T_k$. The Attacker may also have additional constraints, such as a policy of not using more than three bombs on a single target. The feasible choice set for the Attacker is denoted by $\mathcal{A}$.

The interaction between the Defender and the Attacker at the ith target produces a random outcome S_i which takes a value $s_i \in \mathcal{S}_i$. A specific set of outcomes across all targets is denoted by $s = (s_1, \ldots, s_I)$, where $s \in \mathcal{S} = \mathcal{S}_1 \times \cdots \times \mathcal{S}_I$. The Defender's realized utility is $u_D(D, s)$ and the Attacker's realized utility is $u_A(A, s)$. In some applications the utility might also depend upon the actions chosen by the opponent, and then the utility functions for the Defender and Attacker would be written as $u_D(\boldsymbol{D},\boldsymbol{A},\boldsymbol{s})$ and $u_A(\boldsymbol{D},\boldsymbol{A},\boldsymbol{s})$, respectively. The ARA is a bit more complicated, but conceptually it is straightforward.

The Defender seeks the optimal investment $\boldsymbol{D}^* \in \mathcal{D}$. Often she can make plausible conditional independence assumptions that assert that the outcome at the ith target only depends on the total investments by both opponents for that target, and not upon the outcomes at other targets. That assumption would fail if, e.g., one target was a power plant and another target was protected by an electric fence powered by that plant. The conditional independence assumption can be relaxed, at the cost of having to elicit full joint distributions over outcomes at multiple targets, which imposes significant cognitive burden. So this ARA assumes conditional independence, and the Defender need only assess the probabilities $p_D(s_i \mid \boldsymbol{d}_i,\boldsymbol{a}_i)$, where $\boldsymbol{d}_i = (d_{i1},\ldots,d_{iJ})$ and $\boldsymbol{a}_i = (a_{i1},\ldots,a_{iK})$, for $i = 1,\ldots,I$.

The Defender calculates her expected utility for each feasible allocation $\boldsymbol{D}$, conditional on each feasible allocation of the attack, $\boldsymbol{A}$:

$$\psi_D(\boldsymbol{D}\mid\boldsymbol{A}) = \int_S u_D(\boldsymbol{D},\boldsymbol{s}) \prod_{i=1}^{I} p_D(s_i\mid \boldsymbol{d}_i,\boldsymbol{a}_i)d\boldsymbol{s}.$$

If the Defender is also able to assess probabilities $p_D(\boldsymbol{A}\mid\boldsymbol{D})$, reflecting her belief about which attack will be chosen when she selects allocation $\boldsymbol{D}$, then she can compute her unconditional expected utility for each possible defense:

$$\psi_D(\boldsymbol{D}) = \int_A \psi_D(\boldsymbol{D}\mid\boldsymbol{A})\, p_D(\boldsymbol{A}\mid\boldsymbol{D})\,d\boldsymbol{A}. \tag{5.5}$$

She solves for the optimal defense, $\boldsymbol{D}^* = \mathrm{argmax}_{\boldsymbol{D}\in\mathcal{D}}\ \psi_D(\boldsymbol{D})$.

As usual, the trick is to assess $p_D(\boldsymbol{A}\mid\boldsymbol{D})$. Following the method in Section 5.1, the Defender attempts to solve the problem faced by the Attacker. If she knew his utility function $u_A(\boldsymbol{A},\boldsymbol{s})$ and his probabilities for outcomes conditional on defenses and attacks at each target, or $p_A(s_i\mid \boldsymbol{d}_i,\boldsymbol{a}_i)$, then she would calculate his expected utility as

$$\psi_A(\boldsymbol{A}\mid\boldsymbol{D}) = \int_S u_A(\boldsymbol{A},\boldsymbol{s}) \prod_{i=1}^{I} p_A(s_i\mid \boldsymbol{d}_i,\boldsymbol{a}_i)d\boldsymbol{s}.$$

Of course, she does not know his true utilities and probabilities, but she can place subjective distributions over both, and thus can generate random (U_A,P_A). Now she can repeatedly sample and solve

$$\boldsymbol{A}^*(\boldsymbol{D}) = \mathrm{argmax}_{\boldsymbol{A}\in A} \int_S U_A(\boldsymbol{A},\boldsymbol{s}) \prod_{i=1}^{I} P_A(s_i\mid \boldsymbol{d}_i,\boldsymbol{a}_i)d\boldsymbol{s}$$

to find $\hat{p}_D(\boldsymbol{A}\mid\boldsymbol{D})$, her estimate for the probability of the attack which maximizes the Attacker's expected utility. She uses this distribution in (5.5) to find her best feasible allocation.

It is possible to generalize the previous discussion to include cases with multiple attackers, multiple defenders, and multiple targets, all in the same context of a sequential Defend-Attack game. But that topic does not raise new fundamental issues. It is more interesting to generalize to the Defend-Attack-Defend game, in the context of a detailed case study that addresses the issue of spatially related targets.

5.5 Defend-Attack-Defend Games

The sequential Defend-Attack-Defend game assumes that the Defender moves first, choosing a defense. The Attacker observes this choice and responds. Finally, the Defender chooses an action to mitigate the damage from the attack. Clearly, the game could be developed beyond a three-move sequence; it is a dynamic programming problem, where one has fixed planning horizon. Previous work on this game has been done by Brown et al. (2006), Rios and Ríos Insua (2012), Sevillano, Ríos Insua and Rios (2012), and Shan and Zhuang (2014). The game was previously mentioned in Section 4.2 and here it is presented as an extension of the sequential Defend-Attack game.

To fix ideas, suppose a mayor (the Defender) must allocate her police force (and other resources) to protect multiple city districts. Symmetrically, a mob boss directs his criminal henchmen (and other resources) across those districts, trying to maximize his success. Over the course of the game, both the mayor and the mob boss may reshuffle their forces among the districts, and the final score aggregates the results over time for each district. This game is similar to the IED example in Section 2.2.4, where the nodes correspond to districts. The IED analysis assumed that a fixed number of IEDs would be deployed; the present analysis is more flexible with respect to resource allocation.

Suppose the city is divided into I districts, and that the game is played over N days. On each day, the ith district has value v_i^c to the mayor if the cops control it, and value v_i^r to the mayor if the robbers control it, where $v_i^r < v_i^c$. Similarly, on each day, the ith district has value w_i^r to the mob boss if he controls it, and value w_i^c to the mob boss if he does not control it, where $w_i^c < w_i^r$.

Assume that the mayor has J kinds of resources (e.g., policemen, cameras, informants) that she may allocate across the districts. These resources are finite; the jth resource has an upper bound T_j^D, for $j = 1,\ldots,J$. At time t, the mayor assigns an amount $d_{ij}(t)$ of resource j to the ith district, where $0 \le d_{ij}(t)$ and $\sum_i d_{ij}(t) \le T_j^D$. In principle, the magnitude of T_j^D could change over time, if resources are expended or expanded, and there could be interactions among the resources (e.g., some of the policemen could become undercover informants). But this framework avoids so much generality.

Symmetrically, the mob boss controls K resources (e.g., enforcers, arsonists, prostitutes) that he may use in attempts to control the districts. His resources are finite; the kth resource has upper bound T_k^A, and at time t, the mob boss assigns an amount $a_{ik}(t)$ of resource k to district i. Clearly, $0 \le a_{ik}(t)$ and $\sum_i a_{ik}(t) \le T_k^A$.

The mayor and the mob boss allocate their resources, which then interact according to rules that determine the payoffs, and these may be fixed or random. At time $t = 1$, the mayor makes her initial assignment. The mob boss observes this, and at time $t = 2$ makes his initial assignment, and both parties obtain their first payoffs. Then, in alternation, the mayor and the mob boss reallocate their resources, in strategic response to the current situation, and accumulate their payoffs over time. The mayor reallocates when t is odd, and the mob boss when t is even.

Reallocation may entail costs (e.g., administrative overhead, training programs); also, some reallocations may not be possible (e.g., an informant has no useful ties outside his own district). For the mayor, her cost from shifting a unit of resource j from district i to district i' is represented as an element in the $I \times I$ matrix $\boldsymbol{Q}_j$. Similarly, for the mob boss, the reallocation costs for a unit of resource k is given by the $I \times I$ matrix $\boldsymbol{R}_k$.

In this game, when $t = 1, 2, 3$, the interaction between the mayor and the mob boss is a sequential Defend-Attack-Defend game. When $t = 1$, the mayor makes her initial resource allocations to the districts, represented by the $I \times J$ matrix $\boldsymbol{D}_1 = \{d_{ij}(1)\}$. The mob boss observes these assignments, and then, at time step $t = 2$, allocates his resources as the $I \times K$ array $\boldsymbol{A}_2 = \{a_{ik}(2)\}$. At time step $t = 3$, the mayor reallocates her resources, producing the new $I \times J$ matrix $\boldsymbol{D}_3 = \{d_{ij}(3)\}$.

The utilities obtained by the mayor and the mob boss sum their realized values over time for all districts and subtract their reallocation costs. The realized values depend upon the rules that determine which of them controls, or does not control, a district. The details could vary, but this analysis assumes that the mayor has no allocation cost for her initial deployment at time period $t = 1$, and the mob boss has no cost for his initial deployment at $t = 2$.

Figure 5.6 shows the MAID corresponding to this Defend-Attack-Defend game. The notation generalizes to allow indefinitely repeated play, and the rules determining the payoffs could be quite complicated. The chance nodes S_2 and S_3 indicate the random payoffs that the mayor and mob boss receive on the second and third days of the game (without loss of generality, in this formulation, there is no payoff on the first day).

The mayor's perspective on the game is represented by the influence diagram in Figure 5.7, where the attack (or allocation of criminal resources) is the chance node A_2 that reflects the mayor's uncertainty about the mob boss's choice.

In ARA, the mayor must find $p_D(s_2 \mid \boldsymbol{D}_1, \boldsymbol{A}_2)$, $p_D(s_3 \mid \boldsymbol{A}_2, \boldsymbol{D}_3)$, $u_D(\boldsymbol{D}_1, s_2, \boldsymbol{D}_3, s_3)$ and $p_D(\boldsymbol{A}_2 \mid \boldsymbol{D}_1)$. Here, $p_D(s_2 \mid \boldsymbol{D}_1, \boldsymbol{A}_2)$ is her belief about the probability that she controls the districts at time $t = 2$ if her initial resource allocation is the matrix $\boldsymbol{D}_1$ and the mob boss's allocation at time $t = 2$ is $\boldsymbol{A}_2$. Similarly, $p_D(s_3 \mid \boldsymbol{A}_2, \boldsymbol{D}_3)$ is her belief about the probability of controlling each of the districts when $t = 3$, conditional on the reallocation $\boldsymbol{D}_3$ that she makes after observing the mob boss's allocation $\boldsymbol{A}_2$ (this assumes that outcomes are independent over time; it would be violated if, e.g., a burglar was arrested and removed from the game). As usual, the $p_D(\boldsymbol{A}_2 \mid \boldsymbol{D}_1)$ is likely the most difficult to elicit, and is her probability for the allocation the mob boss will make after observing her initial choices. The $u_D(\boldsymbol{D}_1, s_2, \boldsymbol{D}_3, s_3)$ is the total utility she obtains from both her allocation costs and the outcomes for the second

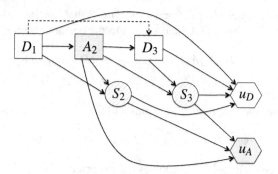

Fig. 5.6 Coupled influence diagrams for the urban security resource allocation problem.

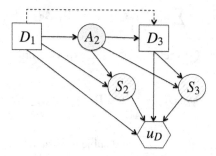

Fig. 5.7 Influence diagram for Mayor's decision problem.

and third time period. Her utility $u_D(\boldsymbol{D}_1, \boldsymbol{s}_2, \boldsymbol{D}_3, \boldsymbol{s}_3)$ sums the utilities $u_D(\boldsymbol{D}_1, \boldsymbol{s}_2)$ and $u_D(\boldsymbol{D}_1, \boldsymbol{D}_3, \boldsymbol{s}_3)$ from each time step.

The mayor should use backwards induction to find her optimal allocation of resources. Specifically:

1. The mayor aggregates the resource expenditures and consequences over all districts and obtains the total utility

$$u_D(\boldsymbol{D}_1, \boldsymbol{s}_2, \boldsymbol{D}_3, \boldsymbol{s}_3) = u_D(\boldsymbol{D}_1, \boldsymbol{s}_2) + u_D(\boldsymbol{D}_1, \boldsymbol{D}_3, \boldsymbol{s}_3)$$

   for every possible scenario $(\boldsymbol{D}_1, \boldsymbol{s}_2, \boldsymbol{D}_3, \boldsymbol{s}_3)$.

2. At chance node $\textcircled{S_3}$ she computes her expected utilities

$$\psi_D(\boldsymbol{D}_1, \boldsymbol{A}_2, \boldsymbol{s}_2, \boldsymbol{D}_3) = u_D(\boldsymbol{D}_1, \boldsymbol{s}_2) + \psi_D(\boldsymbol{D}_1, \boldsymbol{A}_2, \boldsymbol{D}_3),$$

   where

$$\psi_D(\boldsymbol{D}_1, \boldsymbol{A}_2, \boldsymbol{D}_3) = \sum_{s_3} u_D(\boldsymbol{D}_1, \boldsymbol{D}_3, \boldsymbol{s}_3) \, p_D(\boldsymbol{s}_3 \mid \boldsymbol{A}_2, \boldsymbol{D}_3).$$

3. At decision node $\boxed{D_3}$ she finds the actions that maximize her expected utilities, given D_1 and A_2 and subject to the constraints on D_3, as

$$D_3^*(D_1,A_2) = \operatorname{argmax}_{D_3} \psi_D(D_1,A_2,s_2,D_3)$$
$$= \operatorname{argmax}_{D_3} \psi_D(D_1,A_2,D_3).$$

She stores these optimal allocations and the optimal expected utilities

$$\psi_D(D_1,A_2,s_2) = u_D(D_1,s_2) + \psi_D(D_1,A_2,D_3^*).$$

4. At chance node $\textcircled{S_2}$ she computes her expected utilities

$$\psi_D(D_1,A_2) = \sum_{s_2} \psi_D(D_1,A_2,s_2) p_D(s_2 \mid D_1,A_2)$$
$$= \sum_{s_2} u_D(D_1,s_2) p_D(s_2 \mid D_1,A_2) + \psi_D(D_1,A_2,D_3^*).$$

5. At chance node $\textcircled{A}$ she computes the expected utilities for her possible initial decisions

$$\psi_D(D_1) = \sum_{A_2} \psi_D(D_1,A_2) p_D(A_2 \mid D_1).$$

6. Finally, at decision node $\boxed{D_1}$ and subject to the constraints on her initial allocation decision D_1, she finds her optimal initial resource allocation

$$D_1^* = \operatorname{argmax}_{D_1} \psi_D(D_1).$$

This description implies there are a finite number of possible allocations and a finite number of possible outcomes s_2 and s_3, but those assumptions can be relaxed.

As usual, the quality of the mayor's solution depends upon the accuracy of the probabilities and utilities used in the calculation. Among these, usually the most difficult to assess is $p_D(A_2 \mid D_1)$, her distribution on how the mob boss will respond to her initial allocation of resources to districts. If the mob boss has a one-day planning horizon, then the solution is simple, since the mob boss observes her initial allocation and would deploy his resources to maximize his expected gain at time $t = 2$. But if he is more strategic, he may seek a solution that maximizes the sum of his expected gains over a two-day horizon, for times $t = 2$ and $t = 3$. The sequential structure forces the mob boss to be a level-1 thinker, finding his best response against the mayor's observed allocation, and so the mayor must be a level-2 thinker.

Suppose the mayor assumes the mob boss uses a two-day horizon. She assesses her probabilities $p_D(A_2 \mid D_1)$ by solving that problem as if she were he, but with uncertainty over his beliefs and utility that is propagated into her final solution. Figure 5.8 shows an influence diagram that represents the mob boss's perspective on this game, which the mayor uses to find her solution to his game.

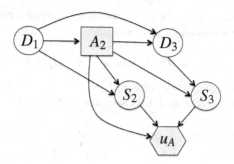

Fig. 5.8 Influence diagram showing the decision problem that the mob boss faces.

In order to maximize his expected utility, the mayor believes the mob boss uses $p_A(s_2 \mid D_1, A_2)$, $p_A(s_3 \mid A_2, D_3)$, $u_A(A_2, s_2, s_3)$, and $p_A(D_3 \mid D_1, A_2)$. These represent, respectively, the mob boss's beliefs about the outcome at time $t = 2$ from choosing allocation A_2 in response to the mayor's choice D_1, the outcome at time $t = 3$ given A_2 and the mayor's reallocation D_3, the utility that the mob boss will receive from outcomes s_2 and s_3 minus the cost of allocation A_2, and the probability of the mayor's reallocation given her initial allocation and the mob boss's allocation.

Since these quantities are not available to the mayor, she models her uncertainty about them through probability distributions (i.e., from her perspective, the probabilities and utilities are random variables). Denote these random quantities by $P_A(s_2 \mid D_1, A_2)$, $P_A(s_3 \mid A_2, D_3)$, $U_A(A_2, s_2, s_3)$, and $P_A(D_3 \mid D_1, A_2)$. She has a subjective joint distribution F over these quantities. She can now propagate her uncertainty using a slightly modified version of the standard influence diagram reduction algorithm (Shachter, 1986) to obtain the mob boss's (random) optimal attack $A_2^*(D_1)$ as a response to each of her possible allocations D_1.

First, she draws a random vector of probabilities and utilities from F. Then, for each D^1, the reduction algorithm is as follows:

1. For each scenario (A_2, s_2, s_3), calculate the random utility $U_A(A_2, s_2, s_3)$.
2. At chance node (S_3) compute the random expected utilities

$$\Psi_A(A_2, s_2, D_3) = \sum_{s_3} U_A(A_2, s_2, s_3) P_A(s_3 \mid A_2, D_3).$$

3. At chance node (D_3) compute the random expected utilities

$$\Psi_A(D_1, A_2, s_2) = \sum_{D_3} \Psi_A(A_2, s_2, D_3) P_A(D_3 \mid D_1, A_2).$$

4. At chance node (S_2) compute the random expected utilities

$$\Psi_A(D_1, A_2) = \sum_{s_2} \Psi_A(D_1, A_2, s_2) P_A(s_2 \mid D_1, A_2).$$

5. At decision node $\boxed{A_2}$ compute the (random) optimal attack as a response to each D_1, taking account of the constraints on the Attacker's actions. Denote the optimal attack by

$$A_2^*(D_1) = \text{argmax}_{A_2}\ \Psi_A(D_1, A_2).$$

This process provides the mayor with $p_D(A_2 \mid D_1) = \mathbb{P}[A_2^*(D_1) = A_2]$, her belief about the probability that the mob boss will make allocation A_2 after observing her allocation D_1.

In order for the mayor to estimate her belief, she may use Monte Carlo sampling. She draws, independently and for $i = 1, \ldots, N$ samples

$$(P_A(s_2 \mid D_1, A_2), P_A(s_3 \mid D_3, s_2), U_A(A_2, s_3), P_A(D_3 \mid D_1, s_2))_i \sim F.$$

For each draw she applies the reduction given in steps (1)-(5) to compute the optimal attack $A_2^{*i}(D_1)$ against each D_1. Then, she uses the approximation

$$\hat{p}_D(A_2 \mid D_1) = \frac{\#\{A_2^{*i}(D_1) : A_2^{*i}(D_1) = A_2\}}{N},$$

where $\#\{\cdot\}$ is the cardinality of the set.

In comparison, the other random quantities described by F are relatively easy to obtain, perhaps from previous history or an informant. However, the last component requires strategic thinking and some kind of recursion, as in Section 2.3.

As an ARA illustration for the case when the mayor is a level-2 thinker, suppose the city has three districts. The ith district provides value v_i to the mayor when she controls it. A district provides no value for the mayor when the mob controls it. Also, suppose that the mayor has two types of resources to allocate: a squad car unit and two foot patrolmen. A feasible allocation by the mayor on the tth day is described by (d_1^t, d_2^t), where $d_1^t = (d_{11}^t, d_{12}^t, d_{13}^t)$ indicates which district has the squad car (i.e., $d_{1i}^t = 1$ if the car is in the ith district, and is otherwise 0), and $d_2^t = (d_{21}^t, d_{22}^t, d_{23}^t)$ indicates how many patrolmen are in each of the three districts. For example, a feasible allocation on the first day is $d_1^1 = (0, 1, 0)$ and $d_2^1 = (1, 0, 1)$, so the squad car is in district 2 and there is one patrolman each in districts 1 and 3.

In contrast, the mob boss has only one type of resource, say burglars, and he manages two of these. A feasible allocation of burglars on the tth day is represented by $a^t = (a_1, a_2, a_3)$ where a_i is the number of burglars in the ith district. There are 18 feasible initial allocations D_1 for the mayor and 6 feasible response allocations A_2 for the mob boss on the second day. The allocation by the mayor on the third day is denoted by D_3.

There are constraints and costs associated with reallocation. Only the squad car may change districts, and that entails a cost of c units; the patrolmen must stay in the district assigned on the first day. For the mayor, the reallocation on the morning of the third day, given the initial allocation (d_1^1, d_2^1), sets d_2^3 equal to d_2^1, while d_1^3 may place the squad car in any district. So there are only 3 feasible allocations D_3. In contrast, the burglars are allowed to move to any district, and since the initial allocation is free and this game has a three-day horizon, the mob boss incurs no relocation costs.

For simplicity, this toy example considers only the payoff on the third day—the role of time is to constrain the mayor's choices on the third day, rather than to also accumulate payoffs on a daily basis. The random variable $S_3 = (S_1, S_2, S_3)$ indicates who has control of each of the three districts at time $t = 3$, after decisions D_1, A_2, and D_3 have been made. Each S_i has two possible outcomes, 0 or 1, where a 1 indicates the mayor has control of the ith district, and a 0 means the mob boss has control.

The probability that the mayor controls district i depends upon the resources that both opponents allocate to that district. To start, assume that these probabilities are common knowledge, and depend upon the allocations as shown in Table 5.1. (The last column, headed (S,2), indicates that the mayor has assigned the squad car and both patrolmen to that district, and the other columns are read similarly.) So, neglecting the time superscripts, each district has the same $\mathbb{P}[S_i = 1 \mid d_{i1}, d_{i2}, a_i]$.

Table 5.1 The mayor's and the moss boss's beliefs about the mayor's probability of controlling a district under each possible combination of allocations.

		Mayor's allocation					
		(0,0)	(0,1)	(0,2)	(S,0)	(S,1)	(S,2)
	0	1.0	1.0	1.0	1.0	1.0	1.0
Mob Boss's allocation	1	0	0.5	0.6	0.7	0.8	0.9
	2	0	0.4	0.5	0.6	0.7	0.8

Since outcomes in districts are independent given the allocations, then

$$\mathbb{P}[(S_3 = (s_1, s_2, s_3) \mid A_2, D_3] = \prod_{i=1}^{3} \mathbb{P}[S_i = s_i \mid d_{i1}, d_{i2}, a_i].$$

The mayor wants to choose D_1 and D_3 so as to maximize her total value (the income from the districts minus the possible cost of relocating the squad car). Similarly, the mob boss wants to place his burglars so as to maximize his payoff. Neither incurs any cost from the initial allocation.

Assuming that the mayor and the mob boss are risk neutral and that each sums the values received from the three districts, then the mayor's utility is

$$u_D(s_3, D_1, D_3) = \sum_{i=1}^{3} v_i s_i - cI(\text{squad car moved}), \qquad (5.6)$$

where $I(\text{squad car moved})$ is an indicator function that takes the value 1 if the squad car is relocated and is otherwise 0. And the mob boss's utility function is

$$u_A(\boldsymbol{s}_3) = \sum_{i=1}^{3} w_i\,(1 - s_i), \qquad (5.7)$$

where (w_1, w_2, w_3) indicates the values that the mob boss receives from each district when he controls it. Neither receives value from a district they do not control.

A traditional game-theoretic solution to this toy problem must assume that the probabilities of controlling each district are commonly known. Also, suppose it is a zero-sum game, with $v_1 = w_1 = 10$, $v_2 = w_2 = 20$ and $v_3 = w_3 = 30$. Finally, suppose the cost of reassigning the squad car is $c = 5$.

Backwards induction computes the game theory solution. First, the mayor must find her best feasible response allocation at $t = 3$:

$$\boldsymbol{D}_3^*(\boldsymbol{D}_1, \boldsymbol{A}_2) = \mathrm{argmax}_{\boldsymbol{D}_3} \sum_{\boldsymbol{s}_3} u_D(\boldsymbol{D}_1, \boldsymbol{D}_3, \boldsymbol{s}_3)\,p(\boldsymbol{s}_3 \mid \boldsymbol{A}_2, \boldsymbol{D}_3). \qquad (5.8)$$

Next, one finds the best response allocation at $t = 2$ for the mob boss, who, knowing $\boldsymbol{D}_3^*(\boldsymbol{D}_1, \boldsymbol{A}_2)$, solves

$$\boldsymbol{A}_2^*(\boldsymbol{D}_1) = \mathrm{argmax}_{\boldsymbol{A}_2} \sum_{\boldsymbol{s}_3} u_A[\boldsymbol{s}_3)p(\boldsymbol{s}_3 \mid \boldsymbol{A}_2, \boldsymbol{D}_3^*(\boldsymbol{D}_1, \boldsymbol{A}_2)]. \qquad (5.9)$$

Finally, one obtains the mayor's optimal allocation at $t = 1$, knowing $\boldsymbol{A}_2^*(\boldsymbol{D}_1)$ and solving

$$\boldsymbol{D}_1^* = \mathrm{argmax}_{\boldsymbol{D}_1} \sum_{\boldsymbol{s}_3} u_D[\boldsymbol{D}_1, \boldsymbol{D}_3^*(\boldsymbol{D}_1, \boldsymbol{A}_2), \boldsymbol{s}_3)\,p(\boldsymbol{s}_3 \mid \boldsymbol{A}_2^*(\boldsymbol{D}_1), \boldsymbol{D}_3^*(\boldsymbol{D}_1, \boldsymbol{A}_2)]. \qquad (5.10)$$

Figure 5.9 shows the mayor's expected payoffs on a normalized 0–1 scale for each of the feasible initial allocations of her resources $\boldsymbol{D}_1$. The two triplets below the bar chart indicate the location of the squad car (top) and the patrolmen (bottom). There is a wide range, so thoughtless allocations can lie far from the optimum.

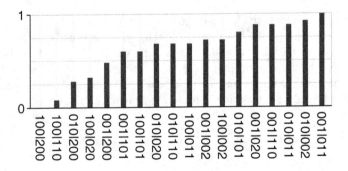

Fig. 5.9 Mayor's normalized expected payoffs for each possible initial allocation $\boldsymbol{D}_1$ under game theory assumptions.

Table 5.2 shows the paths of optimal allocation responses associated with each D_1. One sees that her maximum expected utility decision D_1^* consists of allocating the squad car to her most valuable district, $d_1^1 = (0,0,1)$, and assigning one patrolman to each of her two most valuable districts, $d_2^1 = (0,1,1)$. The mob boss's best response to this is $A_2^*(D_1^*) = (1,0,1)$ or $(0,1,1)$, allocating one of his burglars to the district of highest value and the other to either of the low-value districts (these have equal expected utility). In either case, the mayor responds by leaving the squad where it is, so $D_3^*(D_1^*, A_2^*)$ has $d_1^3 = (0,0,1)$ and $d_2^3 = (0,1,1)$.

Table 5.2 Optimal decision paths for the game tree starting at each of the mayor's feasible initial allocations D_1, along with the expected payoffs to both the mayor and the mob boss.

D_1		$A_2^*(D_1)$	$D_3^*(D_1,A_2^*)$			
d_1	d_2	a	d_1	d_2	u_D	u_A
001	011	101	001	011	44	16
		011	001	011	44	16
010	002	011	010	002	42	18
001	020	101	001	020	41	19
001	110	011	001	110	41	19
010	011	011	010	011	41	19
010	101	011	010	101	39	21
001	002	011	001	002	37	23
100	002	011	010	002	37	18
010	020	101	001	020	36	19
010	110	011	001	110	36	19
100	011	011	010	011	36	19
001	101	011	001	101	34	26
100	101	011	010	101	34	21
001	200	011	001	200	31	29
100	020	101	100	020	27	33
010	200	011	001	200	26	29
100	110	011	010	110	21	34
100	200	011	010	200	19	36

The ARA approach to this toy problem removes the assumption of common knowledge. The mayor does not know either the utilities or the probabilities used by mob boss to solve his decision problem. She must elicit a predictive probability distribution over the mob boss's decision conditional on each of her initial allocations, $p_D(A_2 \mid D_1)$.

To elicit $p_D(A_2 \mid D_1)$, the mayor needs to obtain the following assessments about her beliefs over the mob boss's decision analysis:

- The mayor does not know the mob boss's values w_i for the districts. Suppose the ARA assumes the mayor expresses her uncertainty about those values as $(W_1, W_2, W_3) \sim 60 \times \text{Dirichlet}(1,1,1)$. This leads to a random utility function $U_A(s_3)$, thereby accounting for her lack of common knowledge.

- The game theory solution assumed that the probabilities $p(s_3 \mid A_2, D_3)$ were common knowledge. Instead, ARA allows the mayor and the mob boss to have different beliefs about the probability of controlling a district under each possible combination of allocations. Specifically, suppose the mayor believes the $p_D(S_i = 1 \mid d_{i1}, d_{i2}, a_i)$ are those in Table 5.1, but her beliefs P_A about $p_A(S_i = 1 \mid d_{i1}, d_{i2}, a_i)$, what the mob boss thinks to be the probabilities, are uniform within the intervals given in Table 5.3.

Table 5.3 The mayor's ranges for what she thinks is $p_A(S_i = 1 \mid d_{i1}, d_{i2}, a_i)$, the mob boss's probability that she will control a district under each possible combination of allocations.

		Mayor's allocation					
		(0,0)	(0,1)	(0,2)	(S,0)	(S,1)	(S,2)
Mob	0	1.0	1.0	1.0	1.0	1.0	1.0
	1	0	[0.5, 0.6]	[0.6, 0.7]	[0.7, 0.8]	[0.8, 0.9]	[0.9, 1.0]
	2	0	[0.4, 0.5]	[0.5, 0.6]	[0.6, 0.7]	[0.7, 0.8]	[0.8, 0.9]

For each interval, denote the lower bound by $p_A^{\min}$ and the upper bound by $p_A^{\max}$. Then the random probability she imputes to the mob boss is

$$P_A(S_i \mid d_{i1}, d_{i2}, a_i) = p_A^{\min} + \alpha \left(p_A^{\max} - p_A^{\min} \right) \text{ with } \alpha \sim \text{Unif}(0,1).$$

The uncertainty over α induces uncertainty over p_A to provide P_A. The assessments in Table 5.3 imply that the mayor thinks the mob boss is more pessimistic than she, since the lower bounds to his intervals are her true beliefs.

- If the mayor is a level-2 thinker, then she sees the mob boss as a level-1 thinker who will try to anticipate her decisions. Since he gets to observe her allocation D_1 at $t = 1$, his analysis must concentrate on predicting her response at time $t = 3$, by solving (5.10) The mayor needs then to assess what she thinks are the mob boss's estimates for her u_D and p_D, which he uses to solve her decision problem. The mayor does not know the mob boss's estimate of the value v_i that each district has for her, but suppose she expresses this uncertainty as $(V_1, V_2, V_3) \sim 60 \times \text{Dirichlet}(1,2,3)$. Also, suppose her belief about the mob boss's estimate of her cost $c = 5$ for moving the squad is expressed by $C \sim \text{Unif}(0,5)$, implying that the mob boss tends to underestimate her cost. By using (V_1, V_2, V_3) and C in calculating (5.6), she obtains the random utility function $U_D(s_3, D_1, D_3)$ that represents her belief about what the mob boss thinks is her utility function.

As for her beliefs about the mob boss's estimate of her p_D, it is easy to assume he thinks they share the same beliefs, so $P_D = P_A$.

With this set of illustrative assumptions, the mayor can use simulation to solve the level-2 thinking ARA problem.

The predictive conditional distributions $p_D(\mathbf{A}_2 \mid \mathbf{D}_1)$ are estimated through Monte Carlo simulation with $N = 1000$ replications. For each replication, generate a random set of values (W_1, W_2, W_3), a random set of probabilities $P_A(S_i \mid d_{i1}, d_{i2}, a_i)$, a random set of values (V_1, V_2, V_3), and a random cost C, as described. With these quantities, the mayor can perform the backwards induction in (5.8) and (5.9), producing, for the ith replication, the mob boss's solution $\mathbf{A}_2^{*(i)}(\mathbf{D}_1)$. So the mayor approximates $p_D(\mathbf{A}_2 = a \mid \mathbf{D}_1)$ by $\#\{\mathbf{A}_2^{*(i)}(\mathbf{D}_1) : \mathbf{A}_2^{*(i)}(\mathbf{D}_1) = a\}/N$, for every feasible $\mathbf{D}_1$. This enables her to plug into (5.10) to find her best initial allocation.

Table 5.4 shows the results from this Monte Carlo simulation, where empty cells indicate negligible probability.

Table 5.4 The mayor's Monte Carlo estimates of her predictive conditional distributions for $p_D(\mathbf{A}_2 = a \mid \mathbf{D}_1)$, the probabilities of mob boss's allocations of burglars to city districts given her initial choice.

$\mathbf{A}_2 = (a_1, a_2, a_3)$

$\mathbf{D}_1 = (d^1_{11}, d^1_{21}, d^1_{31} \mid d^1_{12}, d^1_{22}, d^1_{32})$	200	110	020	101	011	002
100|200		0.27	0.03	0.10	0.60	
010|200	0.06	0.09	0.01	0.14	0.69	0.02
001|200	0.08	0.07	0.01	0.13	0.68	0.03
100|110	0.01	0.24	0.02	0.33	0.40	
010|110	0.08	0.11		0.30	0.45	0.06
001|110	0.10	0.13	0.01	0.28	0.41	0.06
100|020	0.02	0.28	0.01	0.54	0.15	
010|020		0.12		0.65	0.20	0.03
001|020		0.13	0.01	0.65	0.19	0.03
100|101		0.48	0.04	0.06	0.23	0.18
010|101	0.09	0.27	0.03	0.11	0.47	0.01
001|101	0.13	0.21	0.03	0.10	0.52	0.01
100|011	0.04	0.47	0.02	0.18	0.10	0.19
010|011	0.02	0.37		0.39	0.20	0.02
001|011	0.01	0.34	0.02	0.43	0.19	0.02
100|002	0.01	0.70	0.02	0.05	0.09	0.13
010|002	0.01	0.63	0.02	0.12	0.22	0.01
001|002	0.01	0.59	0.01	0.13	0.26	

With these probabilities, the mayor uses (5.10) to find that her optimal decision $\mathbf{D}_1^*$ is to allocate the squad car to her most valuable district and one patrolman apiece to each of the other districts, giving her an expected payoff of 44.92. She estimates that under this scenario the mob boss will respond by allocating one burglar to each of her two most valued districts. From Table 5.4, she thinks this response has probability 0.41, but, given her uncertainties, other responses might be the mob boss's best choice, such as assigning one burglar to district 1 and one burglar to district 3, which she thinks has probability 0.28. But even that attack would not lead her to reassign the squad car; that only happens if the mob boss places two burglars in district 2 or one burglar in each of districts 1 and 2. The mayor's analysis leads her to think she will only need to move the squad car with probability 0.14.

Table 5.10 shows the normalized expected payoffs to the mayor from the ARA solution to the game. Interestingly, of the three best ARA solutions ($D_1 = 001|110$, $100|011$ and $010|101$), all have at least one defensive resource in each district. However, only one among the five best game theory solutions has this property, leaving uncovered the district with least value for the mayor. The game theory analysis predicts with certitude that under any initial allocation by the mayor, the mob boss will respond by putting one of his burglars in her district with highest value. In contrast, the ARA takes account of the mayor's uncertainty about the mob boss's probability and utility assessments, leading to a more balanced strategy.

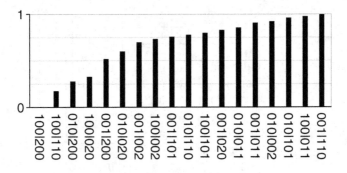

Fig. 5.10 The mayor's expected payoffs (normalized in 0-1) for each feasible initial allocation, as found by ARA.

5.6 Learning

An important aspect of adversarial intelligence is that the opponent learns, and systematically tests the defenses. For example, attackers could both monitor defense activities (e.g., using cameras, or by suborning employees), and probe security (e.g., by measuring reaction times to security breaches). A terrorist organization can test how long it takes for a package to arrive to a destination so that bomb makers are able to tune the timing on the trigger in order to maximize damage. Similarly, defenders may monitor electronic conversations, surveilling for traces of infiltration or threat, as has been done by the U.S. National Security Agency. Other forms of adaptation include the acquisition of new skills, as when the 9/11 terrorists learned to fly airplanes, or discovery of an opponent's strategic reasoning through a combination of feints and reinforcement learning (Kaelbling and Littman, 1996).

In such cases, adversaries could use disclosure to counter an opponent's efforts. Attackers' uncertainty about defender's private information can create opportunities for either defender secrecy or deception (cf. Zhuang, Bier and Alagoz, 2005; Rios and Ríos Insua, 2012; Xu and Zhuang, 2015). Secrecy and deception have been widely studied in military analysis, psychology, computer science, and political science.

Consider a sequential Defend-Attack game where the Defender has information she wants to keep secret. This would arise if there were vulnerabilities in some of the sites she is attempting to protect. The Defender first chooses her defense and, then, having observed it, the Attacker selects his attack. But the Defender's resource allocation decision about protecting different targets might signal the Attacker about site vulnerabilities or the strategic importance of the site to the Defender, which is precisely the kind of information she wants to keep secret.

Assume that the Defender and the Attacker have, respectively, sets $\mathcal{D}$ and $\mathcal{A}$ of possible defenses and attacks. The success level S of an attack is uncertain. Private information is represented by V, whose value is known to the Defender, but not by the Attacker. This information affects the chance of success of an attack, as well as its impact. For both adversaries, the consequences depend, in addition, upon the success of this attack and their own action. Figure 5.11 depicts this situation through a MAID.

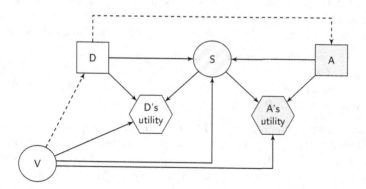

Fig. 5.11 The sequential Defend-Attack model when the Defender has private information.

The utility functions over the consequences for the Defender and the Attacker are, respectively, $u_D(d,s,v)$ and $u_A(a,s,v)$, reflecting that consequences depend also on $V = v$. The arc in the influence diagram from the Defender's decision node to the Attacker's indicates that the Defender's choice is observed by the Attacker. The absence of an arc from Ⓥ to A⃞ indicates that v is not known by the Attacker at the time that he makes his decision.

As before, ARA considers the Defender's problem as one of decision analysis, as shown in Fig. 5.12, with the Attacker's decision node now represented as a random variable to the Defender.

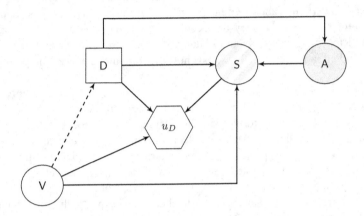

Fig. 5.12 The Defender's decision problem.

Assume the Defender has assessed $p_D(S \mid d,a,v)$ and $u_D(d,s,v)$. She also needs $p_D(A|d)$, which is her assessment of the probability that the Attacker will choose attack a after observing that she has chosen the defense d. Suppose for a moment that she has already assessed $p_D(A \mid d)$. In that case, the Defender can find the defense that maximizes her expected utility through backwards induction as follows:

- At chance node S, compute $\psi_D(d,a,v) = \int u_D(a,s,v)\, p_D(s \mid d,a,v)\, ds$, for each (d,a,v).
- At chance node A, compute $\int \psi_D(d,a,v)\, p_D(a \mid d)\, da$ for each (d,v).
- At decision node D, solve $d^*(v) = \text{argmax}_{d\in\mathcal{D}}\, \psi_D(d,v)$.

Of course, the problem is how the Defender should assess $p_D(A|d)$. The Defender must adopt the perspective of the Attacker and solve his decision problem, as best she can from what she knows. Fig. 5.13 represents the Attacker's decision problem, as seen by the Defender. The arrow from Ⓓ to A⃞ in the influence diagram indicates that the Defender's decision is known to the Attacker when he has to decide. But, since the Attacker does not know the Defender's private information v, his uncertainty is represented through a probability distribution $p_A(V)$, describing the Attacker's beliefs about the Defender's private information. Assume that the Defender thinks the Attacker's is an expected utility maximizer who uses Bayes rule to learn about her private information from observation of her defense decision. Thus, the arrow in the influence diagram from Ⓥ to Ⓓ can be inverted to obtain the

Attacker's posterior beliefs about v: $p_A(V \mid D = d)$. However, it is first necessary to assess $p_A(D \mid v)$.

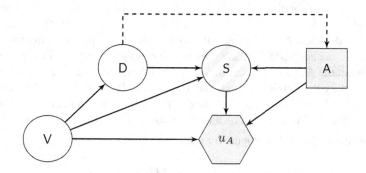

Fig. 5.13 The Defender's analysis of the Attacker's decision problem.

If the Defender knew the Attacker's utility function $u_A(a,s,v)$ and his probabilities $p_A(S \mid d,a,v)$ and $p_A(V \mid d)$, she would be able to anticipate his decision $a^*(d)$ for any $d \in \mathcal{D}$ by computing his expected utility ψ_A. The Defender does not know these quantities, but may be able to model her uncertainty about them through random utilities and probabilities $(U_A(a,s,v), P_A(S \mid d,a,v), P_A(V \mid d)) \sim F$. Then she can propagate her uncertainty through backwards induction to obtain a distribution over the optimal attack $A^*(d)$, as follows:

- At chance node S, compute the random expected utilities $\Psi_A(d,a,v) = \int U_A(a,s,v) \, P_A(s \mid d,a,v) \, ds$, for each (d,a,v).
- At chance node V, compute for each (d,a) the random expected utilities $\Psi_A(d,a) = \int \Psi_A(d,a,v) \, P_A(v \mid d) \, dv$.
- At the decision node A, solve for the optimal attack, which is a random variable, as $A^*(d) = \text{argmax}_{a \in \mathcal{A}} \Psi_A(d,a)$.

Then, the Defender's predictive distribution about the Attacker's choice given any defense choice d is defined by

$$p_D(A = a \mid d) = \mathbb{P}_F \left[a = \text{argmax}_{x \in \mathcal{A}} \Psi_A(d,x) \right], \forall a \in \mathcal{A}.$$

As seen before, the Defender may use Monte Carlo simulation to approximate $p_D(A \mid d)$ by drawing N samples from F, applying the reduction algorithm, and using the corresponding Monte Carlo approximations.

Thus, the elicitation of F allows the Defender to solve her problem of assessing $p_D(A \mid d)$. The Defender may have enough information available to directly assess $P_A(S \mid d,a,v)$ and $U_A(a,s,v)$. However, the assessment of $P_A(V \mid d)$ typically requires a deeper analysis, as it has a strategic component. Specifically, if the Attacker has prior knowledge of V modeled through $p_A(V)$, his posterior belief about V, after he observes $D = d$, becomes:

$$p_A(V = v \mid d) \propto p_A(V = v) \, p_A(D = d \mid v), \tag{5.11}$$

where $p_A(D = d \mid v)$ models the Attacker's probabilistic assessment of what defense she would choose conditional on each possible value of her private information.

The elicitation of $p_A(D \mid v)$ requires an analysis of how the Attacker analyzes the Defender's decision. Assume that he thinks that she is an expected utility maximizer, and that the decision problem she tries to solve is the one shown in Fig. 5.12. Let A^1 represent the Attacker's decision from a level-1 perspective. The Defender elicits her probability distribution G such that $\left(U_D(d,s,v), P_D(S \mid d,a,v), P_D(A^1 \mid d)\right) \sim G$ describes her beliefs about the Attacker's probabilistic assessments of her utilities and probabilities. The distribution allows her to assess $p_A(D \mid v)$ as follows:

- At chance node S, compute for each (d,a,v),

$$\Psi_D(d,a,v) = \int U_D(d,s,v) \, P_D(S = s \mid d,a,v) \, ds.$$

- At chance node A^1, compute for each (d,v),

$$\Psi_D(d,v) = \int \Psi_D(d,a,v) \, P_D(A^1 = a \mid d) \, da.$$

- At decision node D, solve for each v,

$$p_A(D = d \mid v) = \mathbb{P}_G\left[d = \mathrm{argmax}_{x \in \mathcal{D}} \, \Psi_D(x,v)\right], \ \forall d \in \mathcal{D}. \tag{5.12}$$

Since the Attacker's beliefs, as represented by G, are assessed with uncertainty by the Defender, her analysis should produce a distribution over $P_A(D \mid v)$ in (5.12) instead of a point value $p_A(D \mid v)$. Note also that $p_A(V)$ in (5.11) represents the Attacker's prior knowledge about the Defender's private information. As the Defender does not have access to this distribution, she must assess it probabilistically: $P_A(V)$ is the random variable she uses to describe her uncertainty about $p_A(V)$. Then, from the Defender's perspective, the Attacker's updated knowledge about V in (5.11) is represented by

$$P_A(V = v \mid d) \propto P_A(V = v) \, P_A(D = d \mid v).$$

The key difficulty for the Defender at this step is that, in order to obtain $P_A(D \mid v)$, she must elicit her belief about the Attacker's assessment of the probability model she uses to predict his attack, conditional on her chosen defense; i.e., $P_D(A^1 \mid d)$. To address this, she might climb higher up on the ladder of level-k thinking models, analyzing how the Attacker, in his analysis of her decision problem, thinks the Defender will analyze his decision problem, as described in Section 2.2.3. But usually it is difficult to find usable information at higher levels, in which case she can end the hierarchy of introspection with a non-informative distribution over $P_D(A^1 \mid d)$. Now the Defender can compute her best defensive choice. If that choice is sensitive to the reference distribution, then there is critical information that needs to be discovered before she can make a confident decision. She may have to analyze more data or hire more spies.

Chapter 6
A Security Case Study

Much of the previous discussion has focused on simple examples, often toy examples, of ARA. This chapter is an extended case study based upon a real application in railway security. Many of the details have been changed, partly because the client requires confidentiality, and partly because the actual implementation of the study differed from what might have been ideal. Therefore, the data presented is realistic but not actual data. Nonetheless, this chapter provides an example with realistic complexity, and lays out a blueprint for how an ARA of railway security, or security in any similar situation, should be done.

In this problem, the client is a railway service that is concerned about fare evaders, pickpockets, and, perhaps, potential terrorist threats. Fare evasion has significant economic costs to the service. There are two types of fare evaders: those who are casual and opportunistic, and those who collude with others, e.g., by sending text alerts that inform their fellows about which trains and cars are carrying ticket inspectors. Reddy, Kuhls and Lu (2011), Levine, Lu and Reddy (2013), and Sasaki (2014) discuss fare evasion in transit systems, and Ríos Insua et al. (2015) examines it from an ARA perspective. Pickpocketing is even worse than fare evasion. Smith and Clarke (2000) and Troelsen and Barr (2012) describe its modern practice, and issues related to its detection and prevention. When pickpocketing rates are high, there is loss of public confidence and concern that lax enforcement offers opportunity for serious crime, both of which depress ridership. Finally, terrorism is a major threat to railway transportation (cf. Haberfeld and von Hassell, 2009). Some of the worst terrorist attacks of this century have occurred on trains:

- the UNITA attack in Angola on August 10, 2001;
- the Madrid rail bombings on March 11, 2004;
- the three suicide bombers in the London Underground on July 7, 2005;
- the seven bombs on the Suburban Railway in Mumbai on July 11, 2006;
- numerous attacks in Pakistan, especially that of February 17, 2007.

This case study is presented in stages to show how the increasing complexity elaborates the analysis. Initially, the ARA considers reduction in fare evasion at a single station, with one entrance and moderate daily traffic. Later, the analysis generalizes to more serious threats and security at multiple stations. The focus is on fare evasion and pickpocketing, since the railway service has direct and daily responsibility for those issues. Terrorism is a national problem with many stakeholders; nonetheless, this ARA also discusses countermeasures that the client might adopt.

Pickpocketing and fare evasion reduce revenue and damage the public image of the railway service. Seven standard tools that the railways use to combat these threats are shown in Table 6.1. The first four discourage fare evasion, and the last four address pickpocketing (the fourth mitigates both).

Table 6.1 Seven countermeasures used to discourage fare evasion and/or pickpocketing.

	Threat		Features
	Fare Evaders	Pickpockets	
Inspectors	Prevent/Respond	—	Inspect customers, collect fines
Door Guards	Prevent	—	Control access points
Doors	Prevent	—	Secured automatic access doors
Guards	Prevent	Prevent/Respond	Patrol the facility
Patrols	—	Prevent/Respond	Trained guard and security dog
Cameras	—	Prevent	Discourage pickpockets
PR Campaign	—	Prevent	Alert users about pickpockets

Each of these seven countermeasures is already in place, to varying degrees and in different locations. The question for ARA is which countermeasures should receive additional investment, and by how much. As an example of the practical constraints that arise in such problems, Table 6.2 shows, for each countermeasure, the maximum investment railway executives are willing to contemplate at a single station over a planning horizon of one year. The relevant planning period is one year because it is sufficiently long to measure the efficacy of the measures deployed, and because security expenses are budgeted annually.

Table 6.2 The maximum investment that this hypothetical rail station is willing to make in each countermeasure.

Countermeasure	Inspectors	Door Guards	Doors	Guards	Patrols	Cameras
Maximum number of units allowed	4	3	1	2	4	3
Unit cost (in thousands of euros)	40	20	12	25	33	4

In principle, the cost–benefit analysis of countermeasures should depend only upon the available budget, without ceilings on the number of inspectors or door guards. But there are practicalities. The limit on the number of inspectors reflects the train schedule, since on-board inspectors move from station to station. The limit on the number of door guards is driven by the number of doors in the station. The reluctance to install more than one automatic access door is possibly motivated by the physical layout of the station, in which barriers can create chokepoints in pedestrian traffic flow. Similar explanations could be offered for each of the other constraints.

Casual fare evasion is non-strategic, and can be handled within a standard risk analysis paradigm. But fare evaders who collude are strategic, inviting an ARA solution. The case study treats both, and then proceeds to deterrence of pickpockets, showing how ARA can deal simultaneously with all of these threats.

6.1 Casual Fare Evaders

Casual fare evasion entails income loss to the rail service. Measures to reduce fare evasion also entail costs. To maximize revenue, the rail service should aggregate all relevant costs in its utility function and calculate the optimal degree of investment in security measures (Bedford and Cooke, 2001)

This rail service is like many in Europe, in that stations do not have access gates operated by the passenger's ticket; a person can board the train directly. This honor system invites some abuse. To discourage nonpayment, the rail service hires inspectors who ask people to show them a ticket good for that day and that train. Failure to comply leads to a fine, which partially offsets the loss due to evasion and the salaries of the inspectors.

Figure 6.1 shows an influence diagram for this fare evasion problem. The decision node labeled "Countermeasures" refers to the portfolio of actions that may be employed by the railway service.

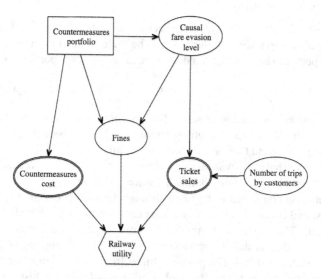

Fig. 6.1 This influence diagram shows the railway services decision problem when considering only the problem of casual fare evaders. Nodes with double circles contain fixed values, conditional on the values in the preceding nodes.

The goal at this stage of the analysis is to maximize the expected utility for the operator by proper selection of the portfolio of countermeasures. Let the components of $x = (x_1, x_2, x_3, x_4)$ denote the number of additional inspectors, door guards, automatic access doors, and guards, respectively. The annual cost for providing an additional inspector is c_1; each door guard costs c_2, each access door costs c_3, and each guard costs c_4. Human resources are budgeted as per-person gross annual salaries. Mechanical resources, such as doors and cameras, are budgeted according to the cost of purchase, installation, and maintenance for one year. The available budget is B, and so the feasible security portfolios x must satisfy

$$c_1 x_1 + c_2 x_2 + c_3 x_3 + c_4 x_4 \leq B,$$
$$x_1 \in \{0,1,2,3,4\},\ x_2 \in \{0,1,2,3\},\ x_3 \{0,1\},\ x_4 \in \{0,1,2\}.$$

For this example, the c_i are given by the third row in Table 6.2, and the available budget is €125,000. There are 50 feasible portfolios; denote this set by $\mathcal{B}$.

To estimate the impact of specific countermeasures on railway loss/profit at one station, assume that the total number of passenger trips per year is N. Let $p_1(x)$ be the proportion of casual fraudsters after security plan x is implemented, and let $q(x_1)$ be the proportion of customers inspected (this depends only upon x_1, the number of inspectors). The railway service expects the following costs and benefits from portfolio x:

- a loss of $c_1 x_1 + c_2 x_2 + c_3 x_3 + c_4 x_4$, for investing in portfolio x;
- expected income from passengers paying fare v, or $N \times v \times [1 - p_1(x)]$;
- expected income from fines paid by caught fraudsters—when the fine is w, this is $N \times w \times p_1(x) q(x_1)$.

Let $x_0 = (0,0,0,0)$. Since $x_0 \in \mathcal{B}$ and is the status quo, then $p_1(x_0)$ is the baseline rate of casual fare evasion; similarly, let q_0 be the baseline rate for passenger trips that are inspected. Then the expected revenue under resource portfolio x is

$$\rho_1(x) = -(c_1 x_1 + c_2 x_2 + c_3 x_3 + c_4 x_4) + N[v(1 - p_1(x)) + w p_1(x) q(x_1)].$$

The railway executives want to select the portfolio of countermeasures such that $\rho_1(x)$ maximizes their expected utility among $x \in \mathcal{B}$. Success requires estimates of N, $p_1(x)$ and $q(x_1)$, and their associated uncertainties.

Estimating the baseline evasion rate $p_1(x_0)$ is simple—for one week, the company can instruct eight inspectors to approach a systematic sample of passengers (e.g., every fifth person encountered as they walk through the train, counting from left to right in the seats, and continuing that tally across cars so that people standing near doors are included). This systematic sampling is unlikely to produce a biased estimate, and provides sufficient data to construct a 95% confidence interval on $p_1(x_0)$ with width approximately 0.004. The protocol differs from the usual practice, in which inspectors have latitude to approach people who appear furtive, so historical data may be misleading about the evasion rate. In actuality, this protocol was not performed; the estimate of $p_1(x_0)$ was based upon the number of evaders who were caught in the previous month and an estimate of the number of people whom the inspectors approached, although this is likely to overestimate the true proportion.

Given an estimate $\hat{p}_1(x_0)$, one can estimate N. The railway service knows N_s, the number of tickets that were sold, so a point estimate is $\hat{N} = N_s/[1 - \hat{p}_1(x_0)]$. There is no closed form expression for the standard error, but it is trivial to simulate it.

Estimating $q(x_1)$, the coverage of inspections, can be done in several ways; e.g., by counting the number of people inspectors approach in a week and dividing by $\hat{N}/52$. But this study used expert judgment. Inspectors provided their subjective estimates, from which the average and variance were calculated.

There was considerable discussion on how to estimate $p_1(x)$, the evasion rate given a specific portfolio. An interesting issue is a positive feedback loop: when many people evade fares, it encourages others to do so too, but this effect is difficult to quantify, and was ignored in the analysis. Ultimately, the decision was made to fit a Cox proportional hazards model with parameter $\boldsymbol{\beta}$ to the ratio of the fare evasion rate under portfolio x to the baseline evasion rate (Cox, 1972). Thus

$$\frac{p_1(x)}{p_1(x_0)} = \exp(\beta_1 x_1 + \beta_2 x_2 + \beta_3 x_3 + \beta_4 x_4)$$

with the constraint that all coefficients must be non-positive, since each countermeasure can only reduce the evasion rate.

Collecting data with which to fit this model would be prohibitively expensive. Instead, the analysts used an expert judgment elicitation protocol. The managers of this station and similar stations, as well as inspectors, guards and other relevant professionals, were separately asked about the expected deterrent effect of each countermeasure for which they had expertise. For example, they were asked to estimate the reduction in evasion from posting one additional door guard. The analysts checked for consistency in their assessments by asking for the expected reduction if two or three door guards were added. As one would anticipate, some judgments diverged. Nonetheless, there was sufficient consensus that it was decided to fit the Cox model to the elicited probability assessments, yielding a point estimate $\hat{\boldsymbol{\beta}}$ with covariance matrix $\hat{\boldsymbol{\Sigma}}$. (In this analysis, the uncertainty in the baseline rate $p_1(x_0)$ was not propagated into the Cox model estimates, but it could have been done at the expense of more computation. The analysts felt that this additional uncertainty was minor compared to the variation in expert opinion.)

Concerning $q(x_1)$, this also required subjective judgment. It is the proportion of passenger trips for which people are asked to present a valid ticket, which depends only upon the number of inspectors. The inspectors who were interviewed felt the increased inspection rate would be sublinear in the number of inspectors (because of light ridership at certain times of day, so that inspectors run out of passengers to approach, and other factors). The expert opinions were averaged to produce $\hat{q}(x_1)$; their uncertainty was estimated by the standard deviation in the opinions.

Finally, from discussions with the railway managers, it was clear that they prefer a sure bet; i.e., the expected value of their utility is less than the utility of their expected value. More specifically, they want to be constant absolute risk averse (CARA) with respect to increasing income (cf. Section 2.2.1). The CARA utility function u_D has the form $u_D[d_1(x)] = 1 - \exp[-\alpha \cdot d_1(x)]$, where $\alpha > 0$ and $d_1(x)$ is the realized overall gain or loss that results from implementing portfolio x, where $\mathbb{E}[d_1(x)] = \rho_1(x)$. To assess the value of the risk coefficient α one may use the probability equivalent method, in which the decision maker chooses between a lottery with a specified probability of winning and a fixed payment (Farquhar, 1984). By varying the probabilities and the amounts it is possible to determine points of indifference, enabling the analyst to estimate α. A very brief and informal version of this method was used in a meeting with senior managers of the railway service.

For each of the 50 feasible portfolios x, one can simulate 10,000 years of operation at the station. The efficacy of the portfolio in each year is inferred from a randomly chosen $p_1(x)$ and $q(x_1)$. The random $p_1(x)$ is obtained from the Cox model where the parameter β is an independent draw $\beta \sim N(\hat{\beta}, \hat{\Sigma})$, but constrained so that all realizations are negative (i.e., one discards draws with one or more positive components, since each countermeasure is believed to reduce fare evasion). Similarly, the inspection coverage $q(x_1)$ in each year is a random draw from the normal distribution with mean and variance equal to the mean and variance in the subjective opinions elicited from the experts—draws that fell outside of $[0,1]$ were discarded. And N, the number of passengers, is a random draw from the normal distribution centered at $\hat{N}$ and with variance obtained by simulation of its standard error. Then, one draws (N_1, N_2, N_3) from a multinomial distribution,

$$(N_1, N_2, N_3) \sim \text{Mult}(N, 1 - p_1(x), p_1(x)q(x_1), p_1(x)(1 - q(x_1))),$$

where N_1 is the number of customers who pay the fare v, N_2 is the number of customers who pay the fine w, and N_3 is the number of customers who pay nothing. With this, one can produce 10,000 estimates of $d_1(x)$. This simulation scheme enables the analyst to calculate the realized utility from each run and average these to estimate the expected utility of portfolio x.

The description of the preceding simulation is close to what was done for the client, with small exceptions. For example, the true ridership N was estimated from a trend analysis of historical data on ticket sales at the station. Also, although the analysts could have propagated uncertainty about the utility parameter α into the analysis, the client preferred that it be treated as a known quantity.

Confidentiality restrictions prevent disclosure of the actual numbers used in the case study, but that is merely a matter of arithmetic. Qualitatively, the optimal portfolio was $(0,1,0,0)$, corresponding to hiring just one door guard. The next best portfolios were $x = (0,0,1,0)$, hiring a security guard, and $x = (0,0,0,1)$, installing an automatic door.

A sensitivity analysis showed that the conclusions depended strongly upon $\hat{p}_1(x_0)$. Repeating this analysis with two somewhat larger values of $\hat{p}_1(x_0)$ found that the two optimal portfolios were (1) the hiring of two door guards and (2) the hiring of one inspector. These results accord with the reasonable conclusion that as the fare evasion rate rises, so should investment in countermeasures.

6.2 Collusion

Some fare evaders collude. For example, the riders on the Paris Metro can pay about €5 a month to join a *mutuelle des fraudeurs*, which insures them against a €60 fine for fare evasion if they are caught (Chu, 2010). And in Melbourne, one can get text alerts about which trains, even which cars, are carrying inspectors (Henriques-Gomes, 2013). The rail service client is very concerned about this emerging trend, and wanted an ARA of the efficacy of the available countermeasures.

This problem is a sequential Defend-Attack game. The colluders observe the security countermeasures *x* taken by the railway station authority and then decide whether or not to attempt fare evasion. Figure 6.2 shows the MAID describing this problem, in which the central chance node indicates the proportion of passengers who decide to collude in fare evasion.

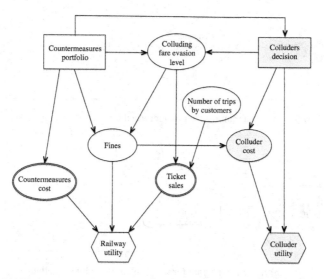

Fig. 6.2 The MAID describing the Defend-Attack game between the rail service and those inclined to collude.

Following the ARA prescription for this class of problems, the rail service decision maker should calculate his belief about how colluders will respond to each of the 50 portfolios, expressing his uncertainty through subjective distributions, and then selecting the portfolio $x \in B$ that maximizes his expected utility, as detailed in Chapter 5. In short, he solves his problem by solving the problem faced by the colluders.

Specifically, denote the current proportion of trips in which someone colludes to avoid payment by $p_2(x_0)$ (i.e., the baseline case with no additional investment) and denote the proportion of trips in which collusion occurs after adopting policy *x* by $p_2(x)$. Then (ignoring casual evasion for the moment) the decision maker wants to maximize his expected utility for $p_2(\boldsymbol{x})$, where

$$p_2(\boldsymbol{x}) = -(c_1x_1 + c_2x_2 + c_3x_3 + c_4x_4) + N[v(1 - p_2(\boldsymbol{x})) + wp_2(\boldsymbol{x})q(x_1)].$$

As before, the c_i terms are the costs of each component of portfolio $\boldsymbol{x}$, N is the number of riders, and $q(x_1)$ is the probability of being caught and fined.

Figure 6.3 shows two influence diagrams. The upper one describes the problem that the railway manager faces. The lower diagram describes the problem that the rational colluders must solve.

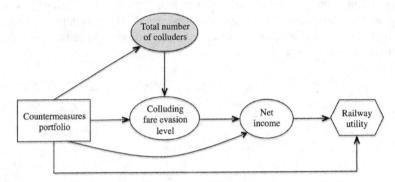

(a) Influence diagram for the problem faced by the rail servic manager.

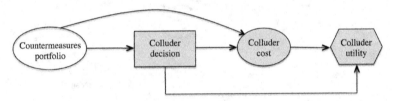

(b) Influence diagram for the problem faced by the colluders.

Fig. 6.3 These influence diagrams show the decision problem from each party's perspective. The rail service manager will solve the colluder's problem in order to optimize his decision.

Recall that $p_2(\boldsymbol{x}_0)$ is the baseline rate of fare evasion by colluders, who purchase insurance and subscribe to text alerts. If portfolio $\boldsymbol{x}$ is used, it can reduce $p_2(\boldsymbol{x}_0)$ in two ways: by increasing the risk of being caught and fined, and by increasing the inconvenience of fare evasion.

The inconvenience of fare evasion is related to the number of door guards, the number of guards, and the installation of a security access gate. For example, a security gate requires either a ticket to operate it or the fraudster must jump the turnstile, which eliminates the less athletic evaders. Similarly, a door guard will not fine someone, but might bar people from entrance until he observes a ticket purchase. The analysts had no reason to think that these disincentives would be different for colluding fare evaders than for casual fare evaders, and thus previous estimates of door guard, guard, and gate effects, derived from the Cox model, may be used. None of these reductions entail adversarial calculation.

In contrast, some fare evaders are discouraged by increased probability of capture, which requires ARA. For each trip, a given fare evader has a certain threshold of risk r_0, $0 \leq r_0 \leq 1$ that he is willing to accept. For example, if a text alert says that an incoming train is carrying an inspector, the passenger will probably decide to purchase a ticket, partly to avoid the delay and embarrassment of being stopped and fined, and possibly so that his insurance rate does not increase. The threshold r_0 for the probability of capture is determined by the individual's personal valuation of the disutility of being caught at that time, and this may change according to mood or circumstance.

ARA cannot know thresholds for individuals, but it may assess a subjective distribution π for the population of colluders. Colluders who were not discouraged by inconvenience will therefore purchase a ticket if the chance of being caught exceeds r_0, where $r_0 \sim \pi$. In this application, a flexible and convenient model for π is the Beta(α, β) distribution, where expert judgment is used to assess α and β.

In this case, the probability of being caught depends only upon the proportion of people approached by inspectors, or $q(x_1)$. The preceding section described an expert elicitation to estimate the increase in inspector coverage as a function of the number of inspectors, and found that it grew sublinearly. Denote those coverage estimates for $x_1 = 0, \ldots, 4$ by $q_0, \ldots, q_4$, respectively. Then the probability that an evader's threshold will be exceeded by an increase in coverage is

$$\mathbb{P}[r_0 \leq q_i] = \int_0^{q_i} \frac{\Gamma(\alpha+\beta)}{\Gamma(\alpha)\Gamma(\beta)} r^{\alpha-1}(1-r)^{\beta-1}\, dr,$$

in which case he would buy a ticket.

The Cox model for casual fare evasion can be rewritten as

$$p_1(\boldsymbol{x}) = \exp(\beta_1 x_1 + \beta_2 x_2 + \beta_3 x_3 + \beta_4 x_4) \times p_1(\boldsymbol{x_0})$$
$$= \exp(\beta_1 x_1) \times \exp(\beta_2 x_2 + \beta_3 x_3 + \beta_4 x_4) \times p_1(\boldsymbol{x_0}).$$

To handle colluding fare evaders, only the term $\exp(\beta_1 x_1)$ should change—unless casual and colluding evaders are differently inconvenienced by door guards, guards, and gates, which seems unlikely, or unless colluding evaders who are not inconvenienced have a different distribution for their risk acceptance threshold than those who are, which also seems unlikely. Therefore, with some confidence, the analysts felt that $\exp(\beta_1 x_1)$ could be replaced by $\mathbb{P}[r_0 \leq q_i]$, so

$$p_2(\boldsymbol{x}) = \mathbb{P}[r_0 \leq q_i] \times \exp(\beta_2 x_2 + \beta_3 x_3 + \beta_4 x_4) \times p_1(\boldsymbol{x_0}),$$

where q_i, $i = 0, \ldots, 4$, is determined by the first component of $\boldsymbol{x}$.

As before, the analyst can now simulate 10,000 years of ridership for each of the 50 portfolios in $\mathcal{B}$. A critical input is $p_2(\boldsymbol{x_0})$, the baseline rate of collusion. For this phase of the analysis, it was estimated in the same way as was done for $p_1(\mathbf{x}_0)$; i.e., the analysis assumed that all fare evaders are colluders, which is unrealistic. However, the rail service requested a pessimistic estimate of the number of colluders, because they fear this is an emerging trend and it wants to plan ahead.

For each run of the simulation, one calculates the expected revenue and finds the CARA utility. Averaging the results gives an estimate of the expected utility for each portfolio.

The primary difference in this simulation, compared to the one for casual fare evaders, is that the estimate of $\mathbb{P}[r_0 \leq q_i]$ is based upon a hierarchical model, to express the uncertainty in the elicited parameter values α and β for the beta distribution. It is cognitively difficult for humans to directly assess a subjective joint distribution over (α, β), but it is relatively easy to place a subjective distribution over the mean μ and standard deviation σ of the risk threshold at which fare evaders decide to buy tickets. Therefore, in each replication, the analysts drew a random mean μ from a normal distribution with mean equal to the average of the experts' judgments on the risk threshold and with standard deviation equal to the standard deviation of their judgments, discarding draws that fall outside of $[0, 1]$. Similarly, they drew a variance σ^2 from a gamma distribution fitted to the expert judgments. Then they solved

$$\mu = \frac{\alpha}{(\alpha + \beta)} \qquad \sigma^2 = \frac{\alpha\beta}{(\alpha + \beta)^2(\alpha + \beta + 1)}$$

for α and β in order to obtain the parameter values used in that replication in the simulation study.

The expert beliefs about the mean and variance of r_0, the risk threshold distribution, are confidential, but the elicitation process is not. Experts were asked to think about several imaginary male evaders: a teenager, an unemployed young adult, a fifty-year-old businessman, and a pensioner. Using these anchors, the experts were asked about each imaginary evader's threshold, and then were asked how changing the gender would alter their belief about the threshold (everyone thought that women had lower risk thresholds) and how changing employment status for the young adult and businessperson would alter their belief (everyone felt that an unemployed person had a higher threshold). These results were weighted by the estimated fraction of passengers with similar demographics to produce an estimate for the mean and standard deviation of r_0. One expert argued that the mean should be at the point at which the cost of the fine times the probability of being caught equals the value of the ticket, but the consensus was that most fare evaders are more risk averse than his reasoning would indicate.

The outcome of the simulation was that the rail service should make no additional investment. The next best choices, with nearly equal expected utilities, were to hire one inspector or one door guard. A sensitivity analysis indicates that these findings are robust when the level of fare evasion is low, but if this rate has been underestimated, then the railway decision makers should make larger investments, generally in the form of hiring more inspectors.

As a final comment, the analysts pointed out to the railway service managers that there is a way to turn the tables on the fare evaders. If colluders are posting text alerts when they see an inspector on a train, then the rail executives could have someone post false text alerts, so that people about to board that train are more motivated to purchase tickets.

6.3 Pickpockets

Suppose now that the railway service wants to defend against pickpockets, and is not concerned with fare evasion. This is a more complex ARA, and the portfolio of countermeasures includes security guards, canine patrols, cameras, and public awareness campaigns that advise passengers to protect their wallets and handbags. This situation is a sequential Defend-Attack game with one Defender and multiple uncoordinated Attackers, as discussed in Section 5.2. Figure 6.4 is a MAID that visualizes the problem. Properly, it should contain one Attacker node for each pickpocket team, but it clarifies both the display and the discussion to consolidate these into one decision node that determines the number of teams.

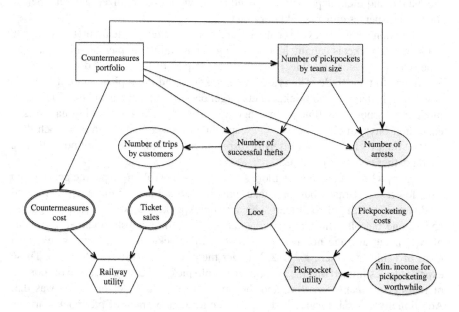

Fig. 6.4 The MAID describing the railway security problem when combating pickpockets.

The previous notation now extends so that portfolio $x = (x_4, \ldots, x_7)$, where x_4 is the number of guards, x_5 is the number of canine patrol units, x_6 is the number of cameras, and x_7 indicates whether or not there is a public awareness campaign to warn passengers about theft ($x_7 = 1$ is a campaign, $x_7 = 0$ means there is not). The feasible security portfolios $x \in \mathcal{B}$ must satisfy:

$$c_4x_4 + c_5x_5 + c_6x_6 + c_7x_7 \leq B,$$
$$x_4, x_5, x_6, x_7 \geq 0,$$
$$x_4 \in \{0,1,2\}, \ x_5 \in \{0,1,2,3,4\}, \ x_6\{0,1,2,3\}, \ x_7 \in \{0,1\}.$$

Table 6.2 shows both the unit costs and the ceilings on investment that railway executives are willing to make in each countermeasure, for all choices except the public awareness campaign. The client had previously sponsored many public awareness campaigns on pickpocket threats, public safety messages, requests to keep the stations tidy, and explanations of the value of inspectors. For the scale of the campaign under consideration, the cost would be €20,000. As before, the available security budget B is €125,000.

To begin the ARA, one first needs to know some facts about pickpockets. Most railway pickpockets operate in small teams among crowds on station platforms; pickpocketing inside a railway car is rare, since there is no escape if things go wrong. Often one confederate jostles a passenger to distract him or her while another performs the theft. If a third team member is present, the thief will inconspicuously pass him the wallet, and then, depending on whether the victim has realized what happened, the team disperses either quickly or quietly.

Often, none of the thieves is caught; rarely is more than one team member caught. When a pickpocket is caught, he is typically fined—failure to pay the fine incurs jail time, but that is unusual. Most pay the fine, as part of the cost of doing business, and return to their trade. The typical fine is €500, and the typical amount stolen is between €100 and €300 (pickpockets target people who appear to be affluent); for modeling purposes, assume the average take is €200. There are additional costs— each thief buys a ticket in order to escape the crime scene by train, and each thief must purchase clothing that disarms suspicion and avoids notice. Suppose that, on average, these costs come to €30 per week.

Larger pickpocket teams are more successful—they rob more people per day and have lower arrest rates, but their take must be divided among more people. Based on discussion with police officers who arrest pickpockets, as well as security guards and canine patrols who work for the rail service, the analysts obtained estimates of $r_i(x_0)$, the average number of successful pickpocket robberies per week for a team of i thieves, where $i = 1, 2, 3$, under the current security portfolio x_0. These experts also estimated that the proportion of solo pickpockets was 0.1, the proportion of two-person teams was 0.6, and the proportion of three-person teams was 0.3. Additionally, the rail service had data on the average number of pickpocket arrests per week at the station of interest; denote this by $z(x_0)$. For obvious reasons, these sensitive numbers are not disclosed.

Local crime experts believe that the annual income a pickpocket needs in order to survive has mean €13,000 and standard deviation €2000. The analysts modeled this through a gamma distribution: each pickpocket's annual requirement is a random draw from the gamma distribution with the specified mean and standard deviation.

When a pickpocket's expected income falls below that draw, he finds other employment (Reilly, Rickman and Witt, 2012).

From the railway service's perspective, pickpockets threaten its income. If too many customers are robbed, people will avoid insecure stations or train lines, and instead carpool or share taxis, thus depressing ticket sales. Senior rail service executives think low rates of theft have little effect, but at some tipping point, there is significant reduction in sales. For this reason, the analysts modeled the sales reduction as a logistic function of the number of robberies. Let $N(t)$ be the expected number of trips per week when there are t thefts per week. Then

$$N(t) = \frac{2(N_0 - N_{min})}{1 + \exp(\theta t)} + N_{min},$$

where N_0 is the average number of weekly trips when there is no crime, N_{min} is the asymptotic limit of the number of trips as the crime rate t goes to infinity (i.e., the number trips taken by people who have no alternative but to ride this train), and $\theta > 0$ is the parameter that controls how quickly ridership declines in response to increased theft.

It is difficult to fit this logistic curve. One knows how many thefts are reported each week, and the weekly ticket sales. But not all thefts are reported, and not all travelers purchase tickets. These figures could be improved by dividing the number of tickets sold by $1 - p(x_0)$, where $p(x_0)$ is an estimate of the proportion of fare evaders, and by dividing the number of reported thefts by $1 - u$, where u is the number of unreported thefts (cf. Moreno and Girón, 1998). Here u might be estimated from interviews with apprehended pickpockets, or from victimization surveys implemented by local governments, or from a poll of passengers. In this application, expert judgment was used to estimate u. The estimate enabled the analysts to specify one point on the logistic curve, but that is not sufficient.

To proceed, one might try to borrow strength from rail and subway services in other cities, or from historical data at this station. Both strategies are problematic, since there may be local differences in risk tolerance and secular trends in ridership rates over time. Instead, the rail service executives decided that if there were 100 more thefts in a week at this station, then ridership would fall by 5% from the current level. That subjective opinion provides a second point on the curve, enabling analysts to fit the logistic response. The study report acknowledged that there is substantial uncertainty about this curve, but the client requested that the analysts not incorporate that uncertainty into the calculation. So the logistic response function was assumed to be known and fixed.

The analysts now need to assess the reduction in theft from each of the four available countermeasures. Adding security guards affects theft in two ways: the guard can apprehend a pickpocket, or the presence of a guard can dissuade pickpockets from making an attempt. The railway service already employs guards, and knows how many arrests these guards make per week. The analysts believe that each additional guard will have a similar, but slightly lower, arrest rate (in the same way that adding additional inspectors was thought to have a sublinear effect). Also, the railway service could study the relationship between the number of guards on a station platform and the number of reported thefts. That evaluation was not performed as part of this ARA, which instead relied upon expert opinion. Let $g(x_4)$ be the expected number of arrests that result from adding x_4 additional guards, and let $g^*(x_4)$ be the probability of successful theft when there are x_4 additional guards.

Similarly, the rail service can use historical data and/or expert opinion to estimate the impact of adding x_5 additional canine patrols. Let $d(x_5)$ and $d^*(x_5)$ be the expected number of apprehensions and the probability of successful theft, respectively, when there are x_5 additional canine units. Railway service employees who had data and expertise provided their estimates of these numbers to the analysts.

In principle, the impact of security cameras could be assessed by cross-station comparisons that relate the number of cameras to the number of reported thefts and the number of arrests. But the way in which cameras are actually used undercuts that methodology. The videotape is studied by guards and canine patrol officers in order to identify pickpockets, so that they will be noticed when they return to the station. Rarely is a video sufficient to enable arrest—almost all arrests involve catching the pickpocket in the act. Since the impact of the cameras is mediated by the number of guards and patrol officers, cross-station comparison is complicated. In this analysis, discussions with currently employed guards and patrol officers were used to estimate $v(x_6)$ and $v^*(x_6)$, their consensus on the number of arrests and the theft probability, respectively, when x_6 new cameras are added. Notably, their judgment was that installing more than two cameras at the station had little impact. Also, they felt that they already knew most of the regular pickpockets by sight, so the marginal value of additional cameras was small.

To assess the effect of a public awareness campaign, the analysts had data. The railway service has periodically run such campaigns over many years, and it was possible to compare the numbers of reported thefts in months when the anti-theft message was being delivered to months in which the campaign delivered public service messages about safety or reminded passengers to dispose of trash properly or assured them that ticket inspectors lead to higher-quality service. Based on that information, the analysts estimated $a(x_7)$ and $a^*(x_7)$, the effects of the presence or absence of a public awareness campaign on the number of arrests and the theft probability, respectively.

Finally, before proceeding to the ARA, the analysts need to estimate the total number of pickpockets who operate at this station during some part of their working day. The rail service provided information on the number of pickpockets who were arrested at this station exactly once, exactly twice, and so forth, during the previous year. Let n be the unknown number of thieves, let K be the total number of arrests, and let k_ℓ be the number of people who have been caught exactly ℓ times. Assuming that all thieves are equally likely to be caught, and that being captured does not alter the chance of future capture, then one can model the situation as drawing K times, with replacement, from a box containing n distinct objects, and then estimating k_0, the number of objects that were not drawn. Good (1953) derived the estimator $k_1^2/2k_2$ for this situation, which he attributed to Alan Turing. In this application, the analysts found that there are about 15 uncaught pickpockets who worked that station during the year (for some portion of their workday).

As usual in sequential Defend-Attack games, the ARA first solves the problem from the perspective of the attacker. A mathematically minded pickpocket would observe the security portfolio x, and either know or soon learn its effect upon his average income. And he would respond strategically, by choosing the action that maximizes his expected utility. In this case, he has the option of seeking other work when pickpocketing becomes insufficiently profitable. A deeper analysis could consider the possibility that he would change the number of people in his team, or move to less protected stations, or work longer hours.

If the elicited estimates are correct, and assuming that the effects of the countermeasures on the number of weekly arrests are additive, then the new weekly average number of arrests is

$$z(x) = z(x_0) + g(x_4) + d(x_5) + v(x_6) + a(x_7).$$

Also, assuming the effects of the countermeasures on theft rate are independent, the new average weekly number of thefts for a team of size i is

$$r_i(x) = g^*(x_4)d^*(x_5)v^*(x_6)a^*(x_7)r_i(x_0),$$

where $r_i(x_0)$ is the average number of thefts by a team of size i under the current, baseline, investment. The portfolio x reduces the pickpocket's income by increasing the number of fines and reducing the number of successful thefts.

The expected number of times a pickpocket is arrested per week is the number of arrests divided by the number of pickpockets, or $z(x)/\hat{n}$. And, if proceeds are divided equally, the expected amount that a pickpocket in a team of size i will receive per week is €$(r_i(x)/i) \times 200$. So, for a pickpocket fine of €500, the expected weekly income of the pickpocket is

$$y_i^* = \frac{1}{i}r_i(x) \times \text{€}200 - \frac{z(x)}{\hat{n}} \times \text{€}500 - \text{€}30.$$

The average annual income is $y_i = 52y_i^*$.

Recall that the annual income a random pickpocket requires has the gamma distribution with mean €13,000 and standard deviation €2000; denote its cumulative distribution function by $F(y)$. Then the probability that a pickpocket in a team of size i must find other work is $F(y_i)$. To find the number of teams of size i, recall that 10% of pickpockets act alone, 60% are teams of two, and the rest are teams of three. And the total number of pickpockets was estimated as $\hat{n}$. Letting T_i be the number of teams of size i, then

$$\hat{n} = T_1 + 2T_2 + 3T_3 = T_1 + 6 \cdot 2 \cdot T_1 + 3 \cdot 3 \cdot T_1,$$

which enables solution for T_1, and, subsequently, for T_2 and T_3.

If the retirement of one member of a team means that the entire team disbands, which seems likely when pickpocketing is barely profitable, then the expected number of remaining solitary pickpockets will be $[1 - F(y_1)]T_1$, the expected number of remaining two-person teams will be $[1 - F(y_2)]^2T_2$, and the expected number of remaining three-person teams will be $[1 - F(y_3)]^3T_3$. So the expected number of weekly thefts, $t(\boldsymbol{x})$, is

$$t(\boldsymbol{x}) = \sum_{i=1}^{3} [1 - F(y_i)]^i T_i r_i(\boldsymbol{x}).$$

In order to calculate the effect of portfolio $\boldsymbol{x}$ on the rail service's income, the analysts are now able to simulate 10,000 years of worth of station operation under each of the 74 feasible portfolios of countermeasures. This was not actually done, since the client wanted the solution that maximized expected utility against both fare evaders and pickpockets. That full analysis is described in the following section. But in principle, for pickpockets alone, one can find the increase in ticket sales over the baseline, $N(\boldsymbol{x}) - N(\boldsymbol{x}_0)$, for each of the 10,000 replicates, and multiply this by v, the average cost of a ticket.

The preceding discussion said little about uncertainty. In fact, for each of the parameters $g(x_4), d(x_5), v(x_6)$ and $g^*(x_4), d^*(x_5), v^*(x_6)$, the analysts estimated the variance in the point estimates from the sample variance in the elicited expert judgments. (There were some exceptions—in a few cases an outlier was removed, and in one case there were very few experts, and so the variance was inflated.) Similarly, for $a(x_7)$ and $a^*(x_7)$, the uncertainty was estimated from the standard errors of the estimates obtained from the data.

6.4 Evaders and Pickpockets

The primary interest of the rail service is to maintain and maximize its revenue. To that purpose, it is considering additional investment in protection against casual fare evaders, colluding fare evaders, and pickpockets, all at the same time. The problem is considerably more complex than the previous analyses in this chapter.

Figure 6.5 is a MAID that visualizes the problem. Light gray nodes correspond to fare evasion through collusion, dark gray nodes refer to theft, and white nodes refer to the rail service, which includes nodes for casual fare evaders, as these are non-strategic. The decision node labeled "Countermeasures" refers to the security portfolio chosen by the railway service to discourage fare evasion and theft.

The notation combines that used earlier. A portfolio is $\boldsymbol{x} = (x_1, \ldots, x_7)$, where the portfolio must satisfy:

$$c_1x_1 + c_2x_2 + c_3x_3 + c_4x_4 + c_5x_5 + c_6x_6 + c_7x_7 \leq B,$$
$$x_1 \in \{0,1,2,3,4\}, \ x_2 \in \{0,1,2,3\}, \ x_3\{0,1\}, \ x_4 \in \{0,1,2\}, \qquad (6.1)$$
$$x_5 \in \{0,1,2,3,4\}, \ x_6\{0,1,2,3\}, \ x_7 \in \{0,1\}.$$

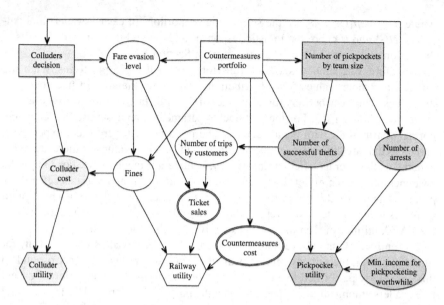

Fig. 6.5 This MAID describes the railway security problem for both fare evaders and pickpockets.

The maximum budget is still $B = €125,000$. Given the other category-specific ceilings listed in Table 6.2, it turns out that the set $\mathcal{B}$ of feasible solutions contains 573 elements, including the baseline case of making no investment, x_0.

Almost all of the elements are in place for a joint solution of the problem of casual and colluding fare evaders, and pickpockets. The missing piece is an estimate of what proportion of the total number of evaders are colluders.

The railway service fears that, in a few years, essentially all evaders will collude, as near-universal ownership of smartphones and evolving social media reduce the technical and normative barriers to such misbehavior. Most experts consulted in this study think that the current proportion of colluders, for this railway service, is small (on the order of 5%). But the rail service executives wanted an ARA that enabled them to plan for the future by examining a range of scenarios.

The analysis of casual evaders described a one-week sampling protocol that would determine the proportion of evaders, although the actual estimate used in the study was based upon administrative records. Such a protocol could be extended so that, when someone is caught, the inspector offers a reduced fine in exchange for information about whether the offender was a colluder, and if so, what kind of collusion took place. In principle, that survey would provide a basis for estimating the proportion of casual evaders $p_1(x_0)$ and the proportion of colluding evaders, $p_2(x_0)$.

Given those numbers, the expected revenue under resource portfolio x is

$$\rho_3(x) = -(c_1x_1 + c_2x_2 + c_3x_3 + c_4x_4 + c_5x_5 + c_6x_6 + c_7x_7) \\ + N(t(x) \times [v(1 - p(x) + wp(x)q(x_1)],$$

where $p(\boldsymbol{x}) = p_1(\boldsymbol{x}) + p_2(\boldsymbol{x})$, the sum of the proportions of casual evaders and colluders, $N(\cdot)$ and $t(\boldsymbol{x})$ are the logistic response function and the number of thefts at the station under portfolio $\boldsymbol{x}$, respectively, as in Section 6.3.

As before, one simulates 10,000 runs for each of the 573 portfolios. Each run samples parameter values from distributions that reflect the elicited uncertainty in the values. The distributions for parameters related to fare evasion were described in Sections 6.1 and 6.2. For the pickpocket parameters, analysts used independent normal distributions with means equal to the point estimates from the experts and variances that are either the sample variance in the elicited values, if there are many experts, or the average elicited uncertainty, if there are few experts. For the public awareness campaign effect, historical data gave the mean and variance. If a draw produced an impossible parameter value (e.g., outcomes in which hiring a new guard increases the theft rate) it was replaced by a valid draw. There was no randomness in the CARA utilities of the rail service executives, by their direction.

Currently, experts believe that fewer than 5% of fare evaders collude. But, for future planning, the railway service wanted to explore a range of scenarios, and so this case study performed five sets of simulations, varying the proportion of casual fare evaders among all fare evaders, so that the proportions were 0, 1/3, 1/2, 2/3, and 1. These results allowed the railway executives to better understand the situation they would face if collusion becomes commonplace.

In all five simulations, the optimal result was to hire one more security guard, which reflects the fact that the security guard is the only asset that can protect against both fare evasion and pickpocketing. However, the realized utilities diminish significantly as the fraction of colluders increases.

The analysts were concerned that the unanimity of the results might be the result of failure to propagate full uncertainty about the logistic response function into the analysis. To assess that, they did a small sensitivity study which varied the parameter θ so that ridership would decline 2% and 10% in response to 100 additional thefts per week. The 2% scenario solution was to a door guard, whereas the solution for the 10% scenario was to hire two security guards.

6.5 Multiple Stations

The rail service in this case study manages more than 100 stations. Most are like the one in the preceding sections—small, with moderate traffic and little crime. But the network contains a hub, which is used by approximately 5 million passengers per year and has a high level of pickpocketing but a moderate rate of fare evasion (possibly because inspectors pass through the hub more frequently than other stations, leading to better enforcement). Also, two stations are known to have high rates of fare evasion, and another has a high rate of pickpocketing.

For purposes of modeling, the analysts used the single station studied previously as the benchmark. The hub and the station with the high theft level were modeled as having pickpocket rates that are twice that of the benchmark. The two stations with high fare evasion rates were modeled as having evasion rates that are twice that of the benchmark. All other stations were treated as being identical to the benchmark. The analysts assumed that 5% of fare evaders colluded, at all stations.

The system-wide security analysis is made much more complex by many additional constraints that reflect budgetary and political realities. Specifically,

- the security budget may not increase by more than €200,000;
- the rail service security budget must increase by at least €120,000;
- the hub must receive at least a €50,000 investment;
- the investment at each of the other stations cannot exceed €30,000;
- one cannot hire more than four new personnel in any job category;
- one cannot install more than four automatic gates;
- one cannot install more than eight cameras;
- either all stations have a public awareness campaign, or none do;
- at most two additional units of each type of countermeasure may be used at a given station, except for the hub.

These requirements greatly complicate the set B of feasible portfolios.

Also, as a minor point, the previous analysis of a single station implied that inspectors are assigned to a single station. From the system perspective, things are a bit more complicated. Inspectors have a home station, and they ride a few stops in each direction away from it. So there is an informal constraint that inspectors should be dispersed among the stations, as it would be duplicative if two home stations were adjacent in the network. But this restriction would be applied administratively, conditional upon the decision to hire an inspector.

There are S stations. The portfolios are $S \times 7$ matrices X, where each row corresponds to the security allocation at a different station. The first row is the hub, the second and third rows are the stations with high fare evasion rates, and the fourth row is the station with a high pickpocketing rate. The remaining $S - 4$ stations are identical to the one in the unitary analysis. Each feasible portfolio must satisfy

$$\text{€}120{,}000 \le \left(\sum_{s=1}^{S} \sum_{k=1}^{6} c_k x_{sk} \right) + c_7 x_7 \le \text{€}200{,}000$$

$$\sum_{k=1}^{6} c_k x_{1k} \ge \text{€}50{,}000$$

$$\sum_{k=1}^{6} c_k x_{sk} \le \text{€}30{,}000,\ s = 2,\ldots,S$$

$$0 \le \sum_{s=1}^{S} x_{sk} \le 4,\ k = 1,\ldots,5,$$

$$0 \le \sum_{s=1}^{S} x_{s6} \le 8$$

$$x_{sk} \le 2\ \ s = 2,\ldots,\ S, k = 1,\ldots,6$$
$$x_{1,7} = x_{2,7} = \cdots = x_{S,7} (= x_7) \in \{0,1\}$$
$$x_{sk}\ \text{nonnegative integer},\ s = 1,\ldots,S,\ k = 1,\ldots,7.$$

The total expected revenue $\rho_4(\boldsymbol{X})$ is just the sum of the expected revenues from each of the stations under that portfolio. One can simulate its distribution for any feasible portfolio assignment $\boldsymbol{X} \in \mathcal{B}$ by drawing parameters according to the previously indicated distributions, separately for each station. For $s = 1,2,3,4$ one modifies some of the distributions in obvious ways to reflect the doubling in the evasion rate or the pickpocket rate.

The challenge in this case study is the exploration of $\mathcal{B}$. Previously, the cardinality of $\mathcal{B}$ was sufficiently small that all elements could be examined, with 10,000 replications under each feasible scenario to estimate the expected utility. But now, the set is too large for enumeration to be practical. Instead, the analysts used a greedy search algorithm with random restarts (Selman, Kautz and Cohen, 1994).

The goal was to find the portfolio that maximizes the expected utility. To do this, a random feasible portfolio $\boldsymbol{X}_1$ was chosen, and 10,000 replications were run in order to estimate its expected utility. Then, one of the allocations in that portfolio was perturbed by a single unit; if that perturbation was in $\mathcal{B}$, then 10,000 more replications were taken at that new portfolio. If the average expected utility was greater, then the new portfolio was adopted as $\boldsymbol{X}_2$ and the perturbation process was repeated. If the average expected utility was the same or less, then the search returned to $\boldsymbol{X}_1$ and a new perturbation was tried. As a minor technical note, the $S-4$ typical stations were handled by simulating one station and multiplying by $S-4$; this greatly reduced the search time, since it was not necessary to explore perturbations that were essentially equivalent, such as hiring a door guard at station 5 versus hiring a door guard at station 6.

The algorithm was randomly restarted 500 times, and found 8 distinct local maxima. By Laplace's Law of Succession, the probability that a new random start would find an undiscovered maximum is $(8+1)/(500+2)$, or approximately 0.02 (Feller, 1968). So the chance of missing the global maximum is even smaller. The time needed for this computation was slightly greater than 15 hours, running on a standard laptop.

Among the eight local maxima, the solution that achieved the largest expected utility is shown in Table 6.3. It also shows the change in expected income, broken out by category. These numbers are not the same numbers found by the analysis described—the agreement with the client precludes that disclosure. But the results are similar to those that were obtained. The Fines column is the expected increase in revenue from fines on pickpockets (but it does not include revenue from fines on apprehended fare evaders, since this portfolio added no inspectors). The Fares column shows the new income from reduced fare evasion. The Theft Avoidance column indicates fares paid by people who otherwise would have avoided use of the rail system for fear of being pickpocketed. Note that Stations 2 and 3 and Stations 5 and 6 are identical, which reflects the modeling assumptions. Stations 5, 6, and 7 are any three of the benchmark stations; the other benchmark stations do not receive investments.

Table 6.3 Optimal portfolio for the bithreat problem in the railway stations. The numbers listed in the columns Fines, Fares, and Theft Avoidance indicate the estimated change from the baseline resulting from implementation of this portfolio.

	x_1	x_2	x_3	x_4	x_5	x_6	x_7	Investment	Fines	Fares	Theft Avoidance
S_1	0	0	0	1	1	0	0	58,000	140,525	156,672	29,211
S_2	0	1	0	0	0	0	0	20,000	—	73,112	—
S_3	0	1	0	0	0	0	0	20,000	—	73,112	—
S_4	0	0	0	1	0	0	0	25,000	54,751	31,736	5,840
S_5	0	0	0	1	0	0	0	25,000	27,118	23,445	2,920
S_6	0	0	0	1	0	0	0	25,000	27,118	23,445	2,920
S_7	0	1	0	0	0	0	0	20,000	—	28,116	—
Total	0	3	0	4	1	0	0	193,000	249,512	409,638	40,891

This final portfolio was the one recommended to the railway service. The analysis is not perfect—many improvements are possible, if not always practical. In some places the account given here differs from the analysis that was actually done, either for reasons of confidentiality or to describe an alternative methodology than the one that was implemented. In terms of the modeling assumptions, the analysts believe that all dominant effects were taken into consideration. A few second-order effects were not; e.g., hiring a door guard or a security guard reduces the rate of fare evasion, which will reduce the revenue generated from fines levied by inspectors.

6.6 Terrorism

A third kind of threat is terrorism. The rail service client did not request an ARA for that threat, and so that analysis was not done, but it is worthwhile to describe some of the issues in framing that problem.

First, the railway service is not responsible for security against terrorism; that is the job of law enforcement and the government, and any steps taken by the railway service must be coordinated with appropriate officials. Nonetheless, the rail service does carry responsibility for responding to foreseeable emergencies, such as fires and passenger injuries. Therefore, each station has an evacuation plan, telephone numbers for the nearest hospitals, and one or more employees trained in first aid—these would all be helpful in the event of an attack. Additionally, the railway service is apprised by law enforcement when credible threats are made against stations, tracks, or cars, and can request an increased police presence.

The first question is whether terrorism in this rail system should be modeled as strategic or non-strategic. Some terrorists are mentally ill, or care about destruction but are indifferent to their choice of target. In those cases, it is probably best to model such attacks as random events. Other terrorists are more deliberative, and these should be modeled as strategic opponents. This is similar to the distinction between casual and colluding fare evaders.

For non-strategic attackers, the railway should use probabilistic risk analysis to assess the danger. It would probably purchase insurance, and the rate charged by the insurance company will depend upon the kinds of security measures that the railway has adopted. This would lead the rail service to perform cost–benefit analyses of specific counterterrorism options, in the context of both the insurance rate and expected reduction in harm.

ARA is more relevant to a strategic opponent. The problem is a sequential Defend-Attack game if the calculation does not include a response plan (e.g., an evacuation protocol or first-aid training), or a Defend-Attack-Defend game, if such response plans are part of the decision making. These cases were discussed in Sections 5.1 and 5.5, respectively. Specifically, this is a game with at least two Defenders (the railway service and the government, and possibly more if there are separate government agencies) and multiple Attackers (unaffiliated or loosely affiliated terrorists).

The general framework for this kind of problem was developed in Section 5.3. The present application is too unspecific to warrant detailed modeling. But it should be noted that, against a strategic opponent, the railway should try to be the second softest target. Rational terrorists will attack the easiest target, while the railway minimizes its security investment (Heal and Kunreuther, 2007).

Chapter 7
Other Issues

The previous chapters laid out the foundation for adversarial risk analysis (ARA). They defined key concepts, developed models and methods, discussed case studies, and placed ARA in the context of both traditional game theory and probabilistic risk analysis. This chapter points forward, to new research challenges. In particular, it addresses issues in modeling complex games, such as computation and improved elicitation, and the identification of novel applications.

7.1 Complex Systems

Real problems are generally quite complex. Opponents need not move either simultaneously or sequentially, but instead will do some of both, and perhaps make decisions according to schedules driven by other factors than the actions of their adversaries. And some games have no clear endpoint; e.g., the competition between the Coca-Cola Company and Pepsico is protracted, with subgames related to product development, business acquisitions, and market share in developing nations. Such scope defies strategic calculation; instead, one must scale up the ARA building blocks so as to provide useful heuristics.

For example, by using ARA techniques for the simultaneous Defend-Attack game (Section 2.1), the sequential Defend-Attack game (Section 5.1), and the Defend-Attack-Defend game (Sections 4.2 and 5.5), one can build models for the Defend-Attack-Defend-Attack scenario considered by Brown, Carlyle and Wood (2008) or the sequential Defend-Attack with Private Information considered by Rios and Ríos Insua (2012) and in Section 5.6. As these games become more elaborate, the subjective beliefs available to the decision maker typically become more vague, and the benefit of the computation is to allow a sensitivity analysis of the solution across a range of assumptions. Sometimes, in complex applications, the most one can gain is knowledge of which choices are clearly bad.

To indicate a situation that is sufficiently complex that useful subjective beliefs are unlikely to be available, consider the MAID in Fig. 7.1. It shows the interaction between two firms, A and B, that compete for market share. Each firm must decide simultaneously how much it will invest in research, and then decide about its production plan. Both firms have private information, e.g., about new products under development and operational costs, and both firms have noisy information about the other's circumstances (and some of that noisy information may be disinformation). Suppose that A is a market leader and B is a follower, so B makes its production decision after A. Earnings depend on the market state and the success of the research investment.

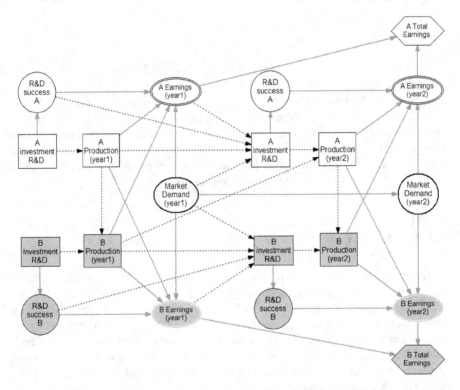

Fig. 7.1 A MAID for complex corporate competition, with leader–follower structure.

The approach to this problem would be based upon the information structure perspective outlined in Section 1.3. One splits the MAID according to the time schedule for the decision making of the firm that the analyst is assisting. One can then distinguish simultaneous versus sequential phases and apply the ideas in Chapters 2 and 4. For example, one starts by solving the influence diagram for firm A, through the standard ID reduction approach (cf. Shachter, 1986). The reduction enables solution for the first A step and generates the influence diagram for B through a modified ID reduction algorithm that takes account of random utilities and random probabilities. Then the analyst solves the problem from B's perspective to obtain the next decision. At various time steps, the analyst needs elements from B's problem in order to solve A's problem, and vice versa. Deciding when to jump from A's problem to B's problem and back is relatively straightforward for human beings, but potentially difficult to encode in an algorithm. The use of relevance and component graphs, described in Koller and Milch (2003), would be helpful in automating this kind of analysis.

Clearly, in an application of this sort, no human is likely to have sharp subjective opinions that enable a confident solution. Instead, the value of ARA in such cases is that it permits the analyst to explore the expected utilities associated with a range of beliefs about utilities, probabilities, and capabilities. Essentially, the analyst can produce solutions for sets of pessimistic, optimistic, and neutral beliefs. The price is that the ARA has to be performed many times, and hence the advantage of automating the calculation.

Beyond the alternation of perspectives and the switching between simulation and optimization, there are other reasons that ARA computation can be challenging. First, realistic games often have many participants; if each individual's decision must take account of the decisions of everyone else, then there can be combinatorial growth in the number of subproblems that must be solved, which does not scale. Second, computation in network games, such as the convoy routing problem studied in Section 2.2.4, generally do not scale well as the number of vertices and edges increase. Third, many games have complex resource constraints. The case study in railway security in Chapter 6 encountered this problem, and greedy hill-climbing was used with random restart, instead of complete exploration of the portfolio space, to find a good solution. Finally, if the game has a spatial aspect, as in hide-and-seek, then computation is generally difficult, especially if the search space has an odd geometry (Alpern and Gal, 2003).

The elephant in the ARA room is elicitation. As in any subjective Bayesian analysis, one needs personal probabilities over the parameters in the problem. Of course, it is well known that experts are overconfident, and that elicitation can entail many biases (cf. O'Hagan et al., 2006). Even worse, in complex adversarial problems it is unlikely that one person has all the relevant expertise to provide informative beliefs. For the example of the two competing firms, one executive might know a lot about the effect of research investment on product development at firm A, but nothing about firm A's market share. Sometimes one can use multidimensional scaling to create a synthetic expert, as in Das, Yang and Banks (2013), and thus estimate the beliefs that someone would have who was expert in all the areas needed to solve the problem, but that approach is largely untested in ARA applications. Nau (1992) and Wang and Bier (2013) both provide detailed discussion of Bayesian elicitation in the context of game theory.

The practical difficulty of elicitation raises the question of robustness. One wants an ARA to be robust to the elicited probabilities, and also to the elicited utilities, the model, and, when available, the data. It is a lot to ask, and no good answer exists.

The best available solution so far is sensitivity analysis. This is the standard practice for simple problems, but complex applications such as counterterrorism are more difficult. Nonetheless, Parnell et al. (2008) recommends it in their National Academy of Sciences review of the U.S. Department of Homeland Security's procedures for assessing bioterrorist threats. And von Winterfeldt and O'Sullivan (2006) performs a systematic sensitivity analysis with respect to elicited probabilities in an event tree concerning Man-Portable Air-Defense Systems (commonly known as MANPADS). Kardes (2007) recommends it when modeling counterterrorism as a robust stochastic game. Complex applications often entail multi-objective decision making, which is its own subfield within the topic of sensitivity analysis (cf. Ríos Insua, 1990).

There are alternatives. Robust Bayesian inference is a methodology that enables the analyst to find the entire set of posterior distributions for a parameter when the prior lies within a ball centered at a specific distribution (cf. Berger and Berliner, 1986). The results are typically expressed in terms of upper and lower probabilities and upper and lower expected utilities. Ríos Insua and Ruggeri (2000) is a review of robust Bayesian inference, but this methodology has yet to be used in ARA.

The only direct application of robustness to ARA is in McLay, Rothschild and Guikema (2012). That elegant paper considers a level-k thinking analysis of the sequential Defend-Attack game in which the Attacker imperfectly observes the decision made by the Defender. The game is modeled through an information structure comprised of several signals and, conditional on the defense choice, there is a specified distribution over the signals—this model was initially proposed by Rothschild, McLay and Guikema (2012). Robustification occurs by setting upper and lower bounds for each of the unknown parameters for which distributions must be elicited, and then calculating the outcome under the worst case combination of upper and lower values. The calculation requires an optimization package, but for the example considered in that paper, the computation was quick.

Robustness tends to overwhelm complex problems. Without special restrictions, the worst case outcome with respect to all possible combinations of upper and lower values for a large number of parameters will be unrealistically pessimistic. But McLay, Rothschild and Guikema (2012) points the way toward a principled means to incorporate robustness into ARA.

ARA is a young field—there are many open research questions, and much to learn about its success in practice. For example:

- Kahneman (2003) and Camerer (2003) and many others have shown that human irrationality shows specific patterns of thinking. How can analysts best take account of that when using ARA?
- In level-k thinking, is it possible to make general statements about the disadvantage of reasoning more than one level higher than the opponent? Are there conditions under which the solution converges as k increases?
- More can be done with formal information structures, disinformation, and information pricing. In many problems that face national security, businesses, or regulatory agencies, these are critical factors.
- In games with more than two players, ARA raises new research questions. Each player may have opinions about what other players think about each of the other players, and these opinions need not agree (Sections 3.6, 4.4.2, 5.2, and 5.3). This is a significant departure from the class of game solutions that depend upon common knowledge.

- The robustification in McLay, Rothschild and Guikema (2012) is an idea that demands further development. One possible extension would base the decision maker's choice upon the distribution of utilities obtained from examining the upper and lower bounds, rather than the worst-case combination.
- Concept uncertainty is a key problem but it is difficult to address (cf. Cox, 2008b; Cardoso and Diniz, 2009). In games with repeated play, one wants to have a procedure for learning how the opponent is reasoning about the problem, based upon game outcomes.

These are some among many of the questions that we believe will challenge ARA research in the next few years.

7.2 Applications

The examples used in this book have mostly been well-studied games, such as *La Relance* and auctions, or variations on Defend-Attack games, or the Colonel Blotto game, or network traversal. The railway case study was novel, but not different; similar decisions are made in many settings, and that discussion merely treated an old and practical problem from an ARA perspective.

But ARA has potential in modern and non-standard situations. For example, consider cybersecurity. Businesses, governments, and private individuals have all fallen victim to cyberattacks, and there are special features of these attacks that set them apart. First, there are large sums of money (or similar assets) that are at risk, and so the effort of deep analysis is warranted, and major defensive investments are already being made. Second, there are many, many different kinds of hackers, and many, many different kinds of targets. Third, the game is asymmetric—attacks are often inexpensive in comparison with the damage they inflict and/or the gains that are won, and some attacks scale easily.

The current approach to risk management for cybersecurity is a patchwork of standards and software and protocols. The International Organization for Standardization recommends use of the ISO/IEC 27000 standard for risk management, which is intended for "any public, private or community enterprise, association, group or individual" (ISO 31000, 2009) and employs standard risk analysis tools such as threat matrices. The National Institute of Standards and Technology produced a guide for protecting information technology systems (Stoneburner, Goguen and Feringa, 2002), and it also is non-adversarial in its analysis, using matrices that sort threats by likelihood and consequence. NATO recently adopted the MAGERIT protocol (Amutio, Candau, and Mañas, 2014), and the government of the United Kingdom has adopted HMG IA Standard No. 1 (National Technical Authority for Information Assurance, 2009). Cox (2008b) criticizes the limitations of such standards: they can assign identical ratings to quantitatively different risks, they can incorrectly assign higher qualitative ratings to quantitatively smaller risks, they can select suboptimal resource allocations, and they refer to ambiguous inputs and outputs. None of these standards explicitly considers the adversarial nature of the threat.

Of course, many cyberattacks are non-strategic, and for these probabilistic risk analysis is appropriate. Most non-strategic attacks upon private individuals involve Nigerian widows asking for help in transferring a sum of money, or Cryptowall holding one's files for ransom (i.e., using malware to encrypt one's computer files, and requiring $500 in bitcoin payment to receive the key), or phishing or keystroke logging to obtain passwords and bank information. These are not crafted to target a specific individual, and so a cost–benefit analysis can determine the value of anti-virus software or other kinds of protection, such as cyberinsurance. Corporations are also subject to non-strategic cyberattacks; the disgruntled employee is, in some sense, a random menace, and company employees may open a mass-mailed malware attachment.

But businesses and governments are also the targets of strategic attacks. The attack upon Sony, to prevent release of *The Interview*, was specific and strategic. Similarly, Edward Snowden's acquisition of National Security Agency data took calculated advantage of flaws in system for authorizing insiders. And there are many other examples, from theft of credit card information at many companies to hacks of the White House email system, files at the U.S. Department of State, and the 2007 denial of service attack upon the Estonian government.

In cybersecurity, many traditional strategies are unreliable. For example, Arce and Sandler (2005) points out that terrorists tend to attack the softest targets. In that case, since security costs money, game theory indicates that one should be the second-softest target. Similarly, when there are many possible targets, such as firms that accept credit cards, then the chance that a specific company will be hacked is low, and thus little investment is required. But the sophistication of cyberattacks grows quickly, and so being the second-softest target is a transient advantage. Also, some cyberattacks are highly scalable; it is possible for an organized hacker to simultaneously target thousands of companies, and thus there is little safety in numbers.

To date, the only ARA of a cybersecurity application is by Couce Vieira, Houmb and Ríos Insua (2014), which focuses upon threats to the oil and gas industry. But a much broader scope is possible and the need is obvious. The next steps might be to analyze other specific cases, such as insider threats or theft of credit card data, and build from there.

A second novel application of ARA is more positive. The idea of modeling one's opponent is fundamentally identical to the idea of modeling one's friend or customer. It may have value in computational advertising, social robotics, and on-line gaming.

In computational advertising, the advertiser wants to present on-line ads to people who will click on them, and ultimately purchase the item on offer. The mechanism for selecting a set of ads for display is complex, and its technical sophistication is growing. It begins when a person types in a few search terms. Advertisers use algorithms to decide how likely it is that this potential customer will be interested in their product, and a rapid auction is held to decide which ads will be shown, and what the placement of those ads on the screen will be (high bidders get the top-listed ads, lower bidders are displayed less prominently, and many bidders lose and their ads are not shown at all). At Google and Yahoo!, the auction is a second-price auction, but there are multipliers on the bids by different companies that reflect the "quality" of the bidder—firms with dodgy reputations or products that may be inappropriate for certain audiences are handicapped or otherwise filtered, and must bid much more to win (Au, 2014).

In this application, it is beneficial to both the advertiser and the potential customer if the advertiser can build a good model for the decision-making process of the customer. In an ideal world, the customer would only be shown advertisements for products that he wants to buy.

A closely related topic is recommender systems. Companies such as Amazon and Pandora track which books or movies a customer browses, or what music the customer likes, and then make recommendations about other books, movies, or music that might also be enjoyed. This requires a smart model for the taste of the individual. A particular challenge is cross-domain recommender systems: If Amazon knows which books a person reads, can it make useful suggestions about what music that person will enjoy? Recently, a few researchers have begun to explore latent factor models in this setting, where some factors reflect the generalized taste of the individual, and other factors are specific to the domain (Au, 2014). The ARA approach would provide a more flexible model for the customer's decision process, which would be helpful in cases for which there is adaptive pricing or repeated contacts with improved recommendations.

Another socially positive application of ARA concerns personal robotics. One example is Siri, an app on Apple's iPhone operating system that acts as a personal assistant. It learns to recognize the iPhone owner's voice and, to some extent, adapt to the owner's previous searches. More ambitiously, there is fast-paced technology that is developing robot cars, automated household help, and even smart houses. If the software in such robots allows them to learn about the habits and preferences of their owners, this could lead to important technological advances in the quality of life. For example, one might build a robot car that reminds the passenger to stop at the laundry or a Roomba Robot (a small vacuum cleaner) that adaptively learns to clean rooms at times convenient to the household or a smart house that notices when someone has fallen asleep watching television and helpfully suggests that she go to bed.

In order to achieve these capabilities, the software needs to let the robot build a model for the decision-making process of its owners. Simple rule-based systems are probably sufficient for many applications, but ARA could help with more sophisticated modeling, e.g., in applications with voice-controlled systems, where "intelligent" conversation and useful suggestions would be a breakthrough. Some ARA work has already been done on social robots. Rázuri, Esteban and Ríos Insua (2013) describes how this approach is used in an edutainment robot designed for children, so that the robot can learn the child's mood and respond constructively. And Esteban and Ríos Insua (2014a) explores how ARA can enable robot analogues to companion pets, which sometimes used by the elderly and people with psychological problems, so that these would seem more empathic, while also adaptively promoting exercise, mood improvement, and other positive activities.

In short, there are many potential new applications for the ARA perspective, both in competitive and cooperative situations. The methodology is often complex, and may pose both cognitive and computational challenges. But this is no different from classical game theory, and not much more difficult that probabilistic risk analysis. Compared to those alternatives, there is good reason to hope that ARA can provide better solutions at comparable cost.

Solutions to Selected Exercises

Chapter 1

1.1 Since the probability of a hurricane given the forecast is f, then Daphne's expected utilities from the two possible decisions are:

$$\psi(E = 1, f) = u(E = 1, H = 1)\, p(H = 1|f) + u(E = 1, H = 0)\, p(H = 0|f)$$
$$= f + w\,(1 - f)$$
$$\psi(E = 0, f) = u(E = 0, H = 1)\, p(H = 1|f) + u(E = 0, H = 0)\, p(H = 0|f)$$
$$= 1 - f.$$

So Daphne will order an evacuation iff $f/(1 - f) > 1 - w$. For example, if $w = 0.8$, she prefers to evacuate iff $f > 0.25$.

1.2 Daphne's utility function is

D	A	S	$u_D(d, a, s)$
0	1	1	0
0	1	0	1
0	0		1
1	1	1	$-c$
1	1	0	$1 - c$
1	0		$1 - c$

where $D = 0$ means she does not install body scanners and $D = 1$ means that she does. Daphne's expected utility for her two possible decisions are $\psi_D(d = 0) = 1 - t_0 v_0$ and $\psi_D(d = 1) = (1 - c) - t_1 v_1$. So $d^* = 1$ iff $t_0 v_0 - t_1 v_1 > c$.

1.3 Given $d = 0$, Apollo's random expected utilities for his two possible decisions are $\Psi_A(a=1, d=0) = y$ and $\Psi_A(a=0, d=0) = x$, and therefore $a^*(d=0) = 1$ iff $y > x$ which Daphne believes to occur with probability $t_0 = 0.3$. And $a^*(d=1) = 1$ iff $y/2 > x$ so $t_1 = 0.15$. Finally, Daphne's maximum expected utility decision is $d^* = 1$ because $t_0 v_0 - t_1 v_1 = 0.3 \times 0.2 - 0.15 \times 0.1 > c = 0.01$.

1.4 The two pure strategy Nash equilibria for this game are (Straight, Swerve) and (Swerve, Straight). To prove this, note that Straight is better than Swerve against Swerve since $W \succ T$; Swerve is better than Straight against Straight since $L \succ C$. Thus, both strategies are best responses to each other. Therefore, together they constitute a Nash equilibrium. This is shown in the game payoff bimatrix as follows:

	Swerve	Straight
Swerve	T , T	L , __W__
Straight	__W__ , L	C , C

These are the only pure strategy Nash equilibria. But a mixed strategy equilibrium also exists.

Multiple Nash equilibria in a game is undesirable, whether one wants to advise a player or predict a choice. The symmetry of this game shows that there is no satisfactory way to reduce the number of equilibria to one. The Nash equilibrium depends upon the assumption that players do not communicate. Thus, if players decide to play a Nash equilibrium in this game but they cannot coordinate, it is conceivable that both players may end up playing Straight, which is the row and column's equilibrium strategy, respectively, in (Straight, Swerve) and (Swerve, Straight). But the pair (Straight, Straight) is not an equilibrium. This not only shows the limited predictive power of the Nash equilibrium concept but also that its use for decision support may be fatal.

1.5 Matching Pennies lacks a pure strategy Nash equilibrium. There is a unique mixed strategy Nash equilibrium in which players choose Heads or Tails with equal probability. To prove this, first note that under this equilibrium, each player's expected payoff is 0. The expected payoff of playing Heads with probability p vs playing Heads with probability $1/2$ is

$$p\left(\frac{1}{2} - \frac{1}{2}\right) + (1-p)\left(-\frac{1}{2} + \frac{1}{2}\right) = 0.$$

Thus, no player has an incentive to choose a strategy with $p \neq 1/2$, given that the opponent plays the equilibrium strategy.

To see that this is the only Nash equilibrium, obtain the players' best response against playing Heads with any probability p such that $0 \leq p \leq 1$. Then, verify that there is only one intersection point.

1.6 Both games are simultaneous. Thus players choose without knowing the other player's choice. It is simple to compare their Nash equilibria with the solutions of a sequential game in which one of the players moves first and the second player chooses after observing the decision. For the Game of Chicken there is a first-mover advantage. If a driver shows his opponent will drive Straight, the best response for second player is Swerve. By changing the game from simultaneous to sequential, one of the two pure strategy equilibria disappears. In simultaneous games of this kind, a similar outcome emerges if a player can be the first to make an irrevocable commitment to a choice.

In contrast, for the Matching Pennies game, there is a second-mover advantage. If one player knows what the other player has chosen, he can always force a win. So Matching Pennies players will try conceal their selection before the simultaneous reveal.

1.7 Let $\$0 \leq x_i \leq \$10$ be the ith group member's contribution to the common pot. Given $x = (x_1, \ldots, x_n)$, the total contribution is $C(x) = \sum_{i=1}^{n} x_i$. The amount obtained by the group is

$$T(x) = 2C(x) + \sum_{i=1}^{n} (10 - x_i) = 10n + C(x).$$

It is maximized when everyone contributes $x_i = \$10$ for a total of $T^* = 20n$.

However, each individual is incentivized to defect from the social optimum. The doubled common pot is divided equally among all group members, regardless their contributions. Thus, if $x_1 = \$0$, but $x_j = \$10$ for $j = 2, \ldots, n$, the first player keeps his $\$10$ and receives $\$20(n-1)/n$. When $n > 2$, the defector ends up with more than $\$20$, the amount he obtains from cooperation.

The advantage of defection leads to an equilibrium in which a dominant strategy for each person is to contribute nothing:

$$u_i(x) = 10 + \frac{2}{n}\left(\sum_{j \neq i} x_j\right) + \left(\frac{2}{n} - 1\right) x_i$$

is maximized at $x_i^* = 0$, regardless of the amounts contributed by everyone else. So a Nash equilibrium is for everyone to contribute nothing.

It is better for everyone if there is no defection. But this is not an equilibrium, since each individual can improve his outcome by defecting. The Nash equilibrium solution concept often produces a socially inferior result: some kind of governance is needed to secure a desirable group outcome.

1.8 Both games have the same two Nash equilibrium pairs, with (Change, Change) better than (SQ, SQ) for both players.

Game I

	Change	Status Quo
Change	10 , 10	−1 , 0
Status Quo	0 , −1	0 , 0

Game II

	Change	Status Quo
Change	10 , 10	−1 , 0
Status Quo	0 , −∞	0 , 0

The two games are nearly identical. From a game-theoretic perspective, (Change, Change) stands out as the solution to both. However, empirical play is very different. Practically all participants play Change in Game I, but many choose SQ in Game II, especially when playing the column's role, fearing the $-\infty$ entry. Game theory cannot explain why participants would choose differently in these two games, assuming rational decision making.

From a decision-analytic perspective, these games are very different. In Game I, players expect each other to choose Change so they also choose Change; they receive a utility of 10 if both choose Change and -1 if (exactly) one of them does not. If a player thinks the other will choose Change with probability p, then Change is preferred if it has higher expected utility than SQ: $10p - (1-p) = 11p - 1 > 0$, which occurs when $p > 1/11$. However, for Game II, the column player prefers Change iff $10p - \infty(1-p) > 0$, which can only occur if $p = 1$. Thus, if the column player thinks there is a chance, no matter how small, that the row player chooses SQ, then the column player will choose SQ. And the row player, afraid that the column player might choose SQ, must choose SQ too, since $0 > -1$. This, in turn, confirms the column player's fear that the row player will not choose Change ($p < 1$).

Chapter 2

2.1 This game has two pure-strategy Nash equilibria: (Concert, Concert) and (Ballet, Ballet). A player using a Nash equilibrium strategy cannot decide which choice to make since both Concert and Ballet are parts of a Nash equilibrium solution.

On the other hand, if the wife believes her husband is non-strategic and she models him as choosing Concert with probability p, then she prefers Ballet iff $p < 2/3$. Similarly, a husband's best move against a non-strategic wife who chooses Ballet with probability q is Concert iff $q < 2/3$. Thus, when both partners share optimistic beliefs about their partner's choice ($p, q < 2/3$) or pessimistic ones ($p, q > 2/3$), they will end up with the worst payoff, $(0,0)$. They avoid this by coordinating their beliefs about each other's choice regarding the Concert ($p > 2/3, q < 2/3$) or the Ballet ($p < 2/3, q > 2/3$).

This game has a third Nash equilibrium, a mixed strategy in which $p = q = 2/3$.

2.2 There are three Nash equilibria: two in pure strategies (Play, Play) and (Work, Work), and one in mixed strategies (1/2 Play + 1/2 Work, 1/2 Play + 1/2 Work). However, (Play, Play) dominates the other two. Moreover, as payoffs are common knowledge, players know that (Play, Play) dominates any other pair of strategies, and, therefore, they can coordinate by expecting each to choose Play, providing the dominant equilibrium.

The maximin (pessimistic criterion) solution is for both players to choose Work, leading to $(1,1)$. The maximax (optimistic criterion) solution recommends both players to choose Play, leading to a payoff of $(2,2)$. The Hurwicz criterion uses a weighted sum of the optimistic and pessimistic criteria. In this game, when both weights are 1/2, there is a tie between Play and Work. This example shows how some solution concepts lead to inferior solutions and, therefore, should not be part of any predictive model for players' moves.

2.3 This is a game of asymmetric information. Apollo is uncertain whether Daphne is stronger or weaker, as represented by her type τ_D, known to her but unknown to him. To compute a Bayes Nash equilibrium (BNE) for this game one must assume a common prior over Daphne's type, given by $\mathbb{P}(\tau_D = S) = \alpha$, which represents Apollo's belief about Daphne's type, and it is common knowledge. Then

- If $\alpha > 2/3$: (Fight, Peace) is the only BNE in pure strategies.
- If $\alpha < 2/3$: (d^*, Fight), where $d^*(S) = \text{Fight}$ and $d^*(W) = \text{Peace}$, is the only BNE in pure strategies.
- If $\alpha = 2/3$: both of the above pairs of strategies are BNE.

It is easy to see that (Fight, Peace) is a BNE when $\alpha > 2/3$: (i) Fight is Daphne's best response against Apollo choosing Peace, regardless of her type, since $u_D(d_0, a_1) = 2 > 0 = u_D(d_1, a_1) \ \forall \ \tau_D$; (ii) Peace is Apollo's best response against Daphne choosing Fight iff $u_A(d_0, a_1) = -1 > -2\,\alpha + (1 - \alpha) = \psi_A(d_0, a_0)$, which occurs when $\alpha > 2/3$.

When $\alpha < 2/3$, (d^*, Fight) is a BNE. Suppose Apollo chooses Fight. If Daphne is stronger (type $\tau_D = S$), then she prefers $d^*(S) = \text{Fight} \succ \text{Peace}$ since $u_D(d_0, a_0, S) = 1 > -1 = u_D(d_1, a_0)$, whereas if she is weaker (type $\tau_D = W$), she prefers $d^*(W) = \text{Peace} \succ \text{Fight}$, since $u_D(d_1, a_0) = -1 > -2 = u_D(d_0, a_0, W)$. Thus, Daphne's strategy d^* is a best response against Apollo choosing Fight. On the other hand, if Daphne chooses d^*, Apollo, who does not know Daphne's type τ_D, has the following expected utilities associated with each of his possible moves:

$$\psi_A(d^*, a_0) = u_A(d^*(S) = d_0, a_0, S) \ \mathbb{P}(\tau_D = S) + u_A(d^*(W) = d_1, a_0) \ \mathbb{P}(\tau_D = W)$$
$$= -2\,\alpha + 2\,(1 - \alpha) = 2 - 4\,\alpha$$
$$\psi_A(d^*, a_1) = u_A(d^*(S) = d_0, a_1) \ \mathbb{P}(\tau_D = S) + u_A(d^*(W) = d_1, a_1) \ \mathbb{P}(\tau_D = W)$$
$$= -1\,\alpha + 0\,(1 - \alpha) = -\alpha.$$

Apollo prefers Fight $\succ$ Peace against d^* iff $\alpha < 2/3$. This proves that when $\alpha < 2/3$, the strategies in (d^*, Fight) are best responses to each other and, therefore, they constitute a BNE.

This argument also shows that if $\alpha = 2/3$, then (Fight, Peace) and (d^*, Fight) are both BNE. In addition, it is possible to prove that these are the only BNE in pure strategies.

2.4 In order to prove that the two pairs of strategies in Exercise 2.3 are BNE it was necessary to assume that Apollo's α is common knowledge. This requires (i) that Apollo discloses what he thinks about her type τ_D to Daphne, and (ii) that Daphne believes him. This implies that it should always be in Apollo's interest to report his true belief, and, therefore, there should be no incentive for him to lie about α. However, if Apollo's true belief is $p_A(\tau_D = S) \in (2/3, 3/4)$, he will be better off reporting "$\alpha < 2/3$," since lying about α ensures that the game will end up at the

BNE (d^*, Fight), which provides him a higher expected payoff than (Fight, Peace), which is the outcome if he truthfully reports α:

$$\psi_A(d^*, a_0) = 2 - 4\alpha > -1 = u_A(d_0, a_1), \quad \forall \alpha < 3/4.$$

On the other hand, if $p_A(\tau_D = S) > 2/3$ it is in Daphne's interest to make Apollo's true beliefs common knowledge, since for $\alpha > 2/3$ the solution (Fight, Peace) becomes the BNE and

$$u_D(d_0, a_1) = 2 > \begin{cases} 1 = u_D(d_0, a_0, S) = u_D(d^*, a_0) & \text{when} \quad \tau_D = S \\ -1 = u_D(d_1, a_0) = u_D(d^*, a_0) & \text{when} \quad \tau_D = W. \end{cases}$$

This suggests that Daphne has incentives to misrepresent her type as stronger, since if Apollo believes this with probability larger than $2/3$, she will be better off.

Thus, assuming Apollo's reported α becomes common knowledge, when his beliefs are $p_A(\tau_D = S) \in (2/3, 3/4)$ there is an incentive for him to lie and for Daphne to make sure Apollo reports his true belief. So, when Apollo reports "$\alpha < 2/3$" Daphne cannot trust this information since it may be a lie which, if believed, will hurt her and help Apollo. Therefore, it is unrealistic to assume that "$\alpha < 2/3$" will become common knowledge.

The solution (Fight, Peace) is the only BNE when $2/3 < \alpha < 3/4$ is common knowledge. In this case, Daphne benefits from knowing Apollo's beliefs α by announcing that she will Fight regardless of her type, because even if she is weaker, Apollo, who believes this to be true only with a probability $\alpha \in (1/4, 1/3)$, will still prefer Peace to Fight, which would have been his optimal response against a known weaker Daphne wanting Fight. However, if Apollo misleads Daphne, making her believe that for him "$\alpha < 2/3$," she would expect Apollo to be better off by choosing Fight instead of Peace which, in turn, sways her to play d^* as her best response. On the other hand, Daphne may not believe Apollo's bluff and promises Fight (whatever her type). This forces him to play Peace if his true belief is $2/3 < p_A(\tau_D = S) < 3/4$. In the end, she knows it is best for Apollo to lie under these circumstances. But if Daphne believes Apollo's lie, his threat of choosing Fight will be credible to her and she would rather play d^*, because she plays Peace when she is weaker. This circular argument makes the BNE unstable. Apollo's reported belief that "$\alpha < 2/3$" cannot realistically become common knowledge since this is not trustworthy information. For a player's belief to become common knowledge, its disclosure by that player must be true-telling incentive-compatible, which is not the case here.

2.5 This is a two-person simultaneous game. Level-1 players only know their own payoffs and ignore their opponent's, who is seen as choosing randomly (i.e., a level-0 player). When players only pay attention to their own payoffs, they identify Defect as their dominant choice, since $20 > 10$ against Cooperate and $0 > -10$ against Defect. Thus, payoffs from choosing Defect are better than from Cooperate regardless of the opponent's move, which makes the pair (Defect, Defect) a Nash equilibrium in dominant strategies.

Level-2 players will not only consider their own payoffs but also their opponent's. This perspective allows the revelation that (Cooperate, Cooperate) dominates (Defect, Defect) since $10 > 0$ for both players. Thus, a player needs at least level-2 thinking to realize that (Defect, Defect) is socially inferior after a joint analysis of the payoff bimatrix. This insight may persuade socially inclined level-2 players to engage with each other and agree to jointly play (Cooperate, Cooperate) instead of acting selfishly and ending up at the Nash equilibrium. However, this social optimum is unstable. To secure an agreement, players will have to change the game's payoff structure, e.g., by signing a binding contract that introduces penalties for non-compliance with the agreed moves. But myopic level-1 players will not even realize the existence of this opportunity for joint benefit.

2.8 There is no dominant choice in this game for any of the players. In order for the wife to predict her husband's choice, she models him as non-strategic. Neither of his two solution concepts use her payoff information to infer anything about how she may choose. Concert would be the husband's choice if he were to use the Laplace criterion, whereas he would choose Concert with probability 1/2 if he were to use the fair coin. The wife can represent her uncertainty about her husband's solution concept by a Bernoulli distribution: the husband will use the Laplace criterion with probability α and the fair coin with probability $1 - \alpha$. Based on this, she predicts that her husband will choose Concert with probability $\alpha + (1/2)(1 - \alpha)$ and Ballet with probability $(1/2)(1 - \alpha)$. If the wife's decision criterion is to maximize her expected payoff, she would choose Ballet over Concert iff her probabilistic belief is $\alpha < 1/3$.

2.9 The wife knows that her husband will choose a Nash equilibrium, but she does not know which one of the three. She knows the solution concept used by him but since this does not specify his final choice, then her uncertainty about his choice is epistemic. Without loss of generality, assume that she believes that he will choose the solution (Concert, Concert) with probability α, (Ballet, Ballet) with probability β, and the mixed strategy with probability $1 - \alpha - \beta$. Then, she believes that her husband will choose Concert with probability $\alpha + (2/3)(1 - \alpha - \beta)$. Given her payoffs, she will choose Ballet iff $2\beta - \alpha > 0$.

Chapter 3

3.1 A first-price sealed-bid auction can be modeled as a simultaneous game. It is a game of incomplete information because bidder's valuations are private. One applies here Harsanyi's BNE concept for solving simultaneous games with incomplete information, where a common prior over the bidders' types is assumed. In this case, the bidders' types correspond to their valuations v_i, which are independent and uniformly distributed on $[0, 1]$ under the common prior assumption.

To find a BNE, first compute the ith bidder's *best response* to the bidding strategies $s_j(v_j) = \alpha_j v_j$ for all bidders $j \neq i$. The probability of winning for the ith bidder when he bids b is

$$\mathbb{P}(b > \alpha_j v_j, \text{for all } j \neq i) = \mathbb{P}(v_j < b/\alpha_j, \text{for all } j \neq i) = \frac{b^{n-1}}{\prod_{j\neq i} \alpha_j}.$$

Then, the ith bidder's expected value for a bid b, given the others bid $\alpha_j v_j$, is $(v_i - b) b^{n-1} / \prod_{j\neq i} \alpha_j$, and the bid b that maximizes his expected value is found by solving the first derivative condition: $(n-1)(v_i - b) b^{n-2} - b^{n-1} = 0$. The ith bidder's best response to $s_j(v_j)$ for $j \neq i$ is

$$s_i^*(v_i) = \frac{n-1}{n} v_i.$$

This proves that the set of bidding strategies $s_i(v_i) = \alpha v_i$ with $\alpha = (n-1)/n$ for all bidders $i = 1, \ldots, n$ is a BNE since each of them is the best response to the others.

As the number of bidders n becomes larger, each person will bid closer to their valuations. Thus, in this type of auctions it is in the interest of the seller, who gets $\frac{n-1}{n} \max_i v_i$, to have as many bidders as possible.

3.2 Let $G_1(v)$ and $G_2(p)$ be Bonnie's independent distributions for Clyde's value V and proportion P with respective supports $a \leq v \leq b$ and $0 \leq p \leq 1$. If $a = 0$, Bonnie's distribution F over Clyde's bid $Y = PV$ can be found as in Equation (3.2). If $a > 0$, Clyde's bid distribution $F(y) = \mathbb{P}[PV \leq y]$ is obtained piecewise through

$$F(y) = \int_a^b \int_0^{y/v} g_1(v) g_2(p) \, dp \, dv = \int_a^b g_1(v) G_2(y/v) \, dv$$

for $0 \leq y \leq a$, so

$$F(y) = \int_a^y \int_0^1 g_1(v) g_2(p) \, dp \, dv + \int_y^b \int_0^{y/v} g_1(v) g_2(p) \, dp \, dv$$

$$= G_1(y) + \int_y^b g_1(v) G_2(y/v) \, dv$$

for $a \leq y \leq b$.

Since the support of G_1 is $[a, b]$, then $G_1(y) = 0$ and its density $g_1(y) = 0$ for all $0 \leq y < a$. Thus,

$$F(y) = G_1(y) + \int_y^b g_1(v) G_2(y/v) \, dv, \quad 0 \leq y \leq b.$$

The density of Clyde's bid is then

$$f(y) = \frac{d}{dy} F(y) = \int_y^b g_1(v) g_2(y/v) v^{-1} \, dv, \quad 0 \leq y \leq b,$$

where $f(y) = \int_a^b g_1(v) g_2(y/v) v^{-1} \, dv$, for $0 \leq y < a$. When

$$G_1(v) = (v - 100)/100, \quad \$100 \le v \le \$200$$
$$G_2(p) = p^2, \quad 0 \le p \le 1,$$

the distribution $F(y)$ of Clyde's bid is

$$F(y) = \begin{cases} \frac{y^2}{20000} & \$0 \le y \le \$100 \\ -\frac{y^2}{20000} + \frac{y}{50} - 1 & \$100 \le y \le \$200 \end{cases}$$

with density

$$f(y) = \begin{cases} \frac{y}{10000} & 0 < y < 100 \\ -\frac{y}{10000} + \frac{1}{50} & 100 < y < 200. \end{cases}$$

Thus, Bonnie's bid that maximizes expected profit is

$$x^* = \mathrm{argmax}_x \, (\$160 - x) F(x),$$

which is the solution of

$$0 = \frac{d}{dx} [(\$160 - x) F(x)] = (\$160 - x) f(x) - F(x) = 3x^2 - 1680x + 84000$$

in $100 < x < 200$. The answer is $x^* = \$103.9$ and represents about 65% of her true value $x_0 = \$160$.

3.3 It is difficult to compute Bonnie's distribution over Clyde's analytically—one should use Monte Carlo simulation. Draw n independent samples from both P and V, so $\{p_i\}_{i=1}^n \sim \mathrm{Beta}(20, 10)$ and $\{v_i\}_{i=1}^n \sim \mathrm{Tri}(100, 150, 200)$. A sample from Clyde's bid $Y = PV$ is then generated by multiplying $p_i v_i = y_i$, for $i = 1, \ldots, N$, and its probability distribution is then approximated by the empirical distribution $F(y) = \mathbb{P}[PV \le y] \approx \sum_{i=1}^N \mathrm{I}\{y_i < y\}/N$, where $\mathrm{I}\{\cdot\}$ is the indicator function.

Bonnie's expected profit for each of her bids $\psi(x)$ can be approximated by $(\$200 - x) \hat{F}(x)$. The $\psi(x)$ is continuous in x, since it is the product of two continuous functions $(\$200 - x)$ and $F(x)$, and has a maximum between $x = 0$ and $x = \$200$ since $\psi(0) = \psi(200) = 0$ and $\psi(x) > 0$ in $0 < x < 200$.

To find the maximum of $\psi(x)$, one may use the well-known golden section search. The procedure starts with the interval $[0, 200]$, and iteratively computes a series of shrinking, nested intervals that converge to the maximum. In this case, with $n = 20,000$ and a tolerance of 0.001, it finds, after 26 iterations, that Bonnie's optimal bid is $x^* \approx \$121.96$, which is about 60% of her valuation $x_0 = \$200$.

3.4 If F is continuous with $F(x_0) > 0$, there exists $\varepsilon > 0$ and $x_0 - \varepsilon < x^- < x_0$ such that $F(x^-) > 0$. Then, Bonnie's expected profit from bidding $x^- < x_0$ is

$$u_B(x^-) = (x_0 - x^-) F(x^-) > 0 \ge (x_0 - x) F(x) = u_B(x),$$

for any $x \ge x_0$. In particular, $u_B(x^*) \ge u_B(x^-) > u_B(x)$ for all $x \ge x_0$. So, the bid that maximizes Bonnie's expected profit must be $x^* < x_0$.

3.5 The optimal bid in a sealed-bid first-price auction is $x^* = \text{argmax}_x (x_0 - x) F(x)$, where x_0 represents the value of the item to the bidder, and $F(x) = \mathbb{P}[Y_M < x]$ is her probabilistic assessment that she will win the auction with a bid of x, which occurs when her bid is higher than the maximum bid Y_M of the others.

In a Dutch auction, if the price has dropped below x_0 and no one has spoken out yet, the conditional probability of winning if one waits to bid until $x \leq b$ is $\mathbb{P}[Y_M < x \mid Y_M < b] = F(x)/F(b)$. Thus, at any time a price $b \leq x_0$ is offered by the seller, the bidder should wait for a price $x \leq b$ that maximizes her expected profit $(x_0 - x) F(x)/F(b)$. So holding out for x^* is the optimal strategy here, since $F(b)$ in the denominator is a constant that does not change when the optimum is reached.

In particular, for any $x^* < b < x_0$, the bidder must decide whether to accept b and get a sure profit of $x_0 - b > 0$, or wait until x^*, and then receive the extra profit of $(x_0 - x^*) - (x_0 - b) = b - x^*$ with the risk of losing the auction. The bidder believes the chance of losing has probability $1 - F(x^*)/F(b)$. For example, if $x_0 = \$1,000$, $x^* = \$500$ and $b = \$600$, the bidder must decide between getting $400 for sure or gamble on getting \$100 more with probability $F(x^*)/F(b)$. Normatively, the gamble is preferred when its expected profit is higher than the sure profit: $(x_0 - x^*)F(x^*)/F(b) > (x_0 - b)$ iff $(x_0 - x^*)F(x^*) > (x_0 - b)F(b)$, which is always the case since x^* is the optimal bid.

We have seen that the Dutch auction is formally equivalent to the sealed-bid first-price auction. However, empirically, these two auctions can be quite different. In practice, bidders may not make coherent decisions, consistent with normative theories. For example, when confronted with the Dutch auction choice, many people prefer the sure profit of $400, rather than the gamble for \$100 more, which would be their normative optimal decision. Descriptively, many people deviate from expected utility and undervalue lotteries with higher expected utilities than guaranteed positive consequences. Thus, one expects higher bids in a Dutch auction.

3.6 From the auctioneer's perspective, Bonnie's value for the auctioned item is $x_0 > \$10M$ and her estimate of Clyde's value is $\hat{y}_0 < \$1M$. If the auctioneer is right, Bonnie's bid in a sealed-bid first-price auction would be $x^* \leq \$1M$. To see this, notice that Bonnie's distribution on Clyde's bid will be such that $F(x) = 1$ for all $x > \hat{y}_0$, since she does not expect Clyde to bid y higher than his valuation $\hat{y}_0$ and, therefore, she will win with any bid $x > \hat{y}_0$. Then, the bid x that maximizes her expected profit $(x_0 - x) F(x)$ will be $x^* \leq \hat{y}_0 \leq \$1M$. Similarly, the auctioneer thinks that Clyde will also bid $y^* \leq \$1M$, so the auctioneer obtains no more than $\max(x^*, y^*) \leq \$1M$.

In contrast, in a second-price auction one expects revenue of at least $\min(x_0, y_0) > \$10M$ (the second highest of the bidders' true valuations), where the highest sealed-bid wins but pays the second highest bid, assuming bidders correctly bid their true valuations. Similarly, the auctioneer is better off with an open-cry ascending (or English) auction, where buyers increase their bids until no one will bid more. Again, the winner is expected to be the bidder with the highest valuation and the revenue to be the second highest valuation, which in this case is at least $10M.

3.7 This is a simultaneous two-person auction game of complete and symmetric information. Bidders know each other's valuations, there is no need for subjective probabilities to represent their respective beliefs: the prize (a $100 bill) is worth the same to both and each knows that the other's valuation is this *common value*.

To find the Nash equilibrium of this game, perform a best response analysis. If Bonnie bids x and Clyde bids y, Bonnie will get 0 if her bid loses $(x < y)$ and get $100 - x$ if she wins $(x > y)$. If she were to know that her opponent's bid is $y \leq \$99.98$, she would maximize her profit by bidding $x^*(y) = y + 0.01$. If $y = \$99.99$, $x^*(y) = y$ in order to get a tie and have the opportunity to win a penny from a fair coin tossing (tie-breaking rule). Finally, if $y = \$100$, any bid $x^*(y) \leq 100$ will be optimal with a $0 profit, since a bid x higher than $100 will lead to a loss. She assumes, reasonably, that no bid will be higher than the prize value of $100. So the game has two Nash equilibria: $(x^*, y^*) = (\$100, \$100)$ and $(\$99.99, \$99.99)$, with the equilibrium at $100 dominated by the one at $99.99, which yields an expected profit of half a penny to each bidder.

On the other hand, suppose she knows that in real auctions like this the probability of bids equal to $100 or $99.99 is quite low. Specifically, bids seem to be uniformly distributed between $60 and $100. The ARA approach facilitates finding the optimal bid by solving

$$x^* = \text{argmax}_{0 \leq x \leq 100} (\$100 - x) \, \mathbb{P}[U < x] = 80,$$

where $U \sim \text{Unif}(\$60, \$100)$.

3.8 Let Y and Z represent the bids of the other two participants, which are supposed to be independent and uniformly distributed between $70 and $100. Under the ARA approach, the optimal bid would be

$$x^* = \text{argmax}_{0 \leq x \leq 100} (\$100 - x) \, \mathbb{P}[\max\{Y, Z\} < x]$$
$$= \text{argmax}_{70 \leq x \leq 100} (100 - x)(x - 70)^2.$$

Setting the derivative equal to zero gives

$$0 = \frac{d}{dx} \left[(100 - x)(x - 70)^2 \right] = 3x^2 - 480x + 18900$$

and solving for x in $70 \leq x \leq 100$ finds $x^* = \$90$.

Chapter 4

4.1 This problem is a sequential Defend-Attack game. The Attacker's best response to a defense with x units of protective resources is to attack iff $x < (v/c_A) - 1)$, because

$$u_A(x, \text{ attack}) = -c_A \times x/(x+1) + (v - c_A) \times 1/(x+1) > 0,$$

where 0 is the utility received from not mounting an attack. Knowing the Attacker's solution, the Defender should find the resource investment x^* that will maximize her expected payoff

$$u_D(x) = \begin{cases} v\frac{x}{(x+1)} - c_D x, & x < (v/c_A) - 1 \\ v - x c_D. & x \geq (v/c_A) - 1 \end{cases}$$

Solving for the optimal defense allocation gives

$$x^* = \begin{cases} \sqrt{v/c_D} - 1, & \text{if } c_D > 4c_A^2/v \\ (v/c_A) - 1, & \text{if } c_D < 4c_A^2/v. \end{cases}$$

4.2 This problem is a zero-sum sequential defend-attack game. Let $d_i \geq 0$ be the Defender's (percentage of) resources allocated to Site i. Then, $d = (d_1, d_2)$, where $d_1 + d_2 = 1$, represents all feasible defense allocations. The Attacker decides which site to attack, with $a_i = 1$ if he attacks site i and $a_i = 0$ if he does not. Thus, $a = (a_1, a_2)$ with $a_1 + a_2 \leq 1$ and $a_i \in \{0, 1\}$ represent the attack alternatives. The success probability of an attack on site i is

$$p(S_i = 1 | d_i, a_i) = (1 - d_i) a_i,$$

where S_i is a Bernoulli random variable representing whether the defender loses ($S_i = 1$) site i or not ($S_i = 0$).

A risk-neutral Defender will minimize her expected loss. If she gives values v_1 and v_2 to each site, her loss would be $\ell_D(S_1, S_2) = v_1 S_1 + v_2 S_2$. Therefore, her expected loss is

$$\tilde{\Psi}_D(d, a) = v_1 (1 - d_1) a_1 + v_2 (1 - d_2) a_2,$$

for all feasible alternatives d and a of the Defender and the Attacker, respectively.

Since the game is zero-sum, the Defender believes the Attacker will seek to maximize her expected loss. So she predicts the Attacker's best response to any of her feasible allocations:

$$a^*(d) = \begin{cases} (a_1 = 1, a_2 = 0) & \text{iff} \quad d_1 < v_1/(v_1 + v_2) \\ (a_1 = 0, a_2 = 1) & \text{iff} \quad d_1 > v_1/(v_1 + v_2) \\ \text{indifferent} & \text{iff} \quad d_1 = v_1/(v_1 + v_2). \end{cases}$$

The Defender's optimal allocation is then obtained by solving

$$d^* = \arg\min_d \tilde{\Psi}_D(d) = \tilde{\Psi}_D(d, a^*(d)) = \begin{cases} v_1(1 - d_1) & \text{iff} \quad d_1 \leq v_1/(v_1 + v_2) \\ v_2(1 - d_2) & \text{iff} \quad d_1 \geq v_1/(v_1 + v_2) \end{cases}$$

with the Defender allocating $d_1^* = v_1/(v_1 + v_2)$ (percentage of) resources to the first site and $d_2^* = v_2/(v_1 + v_2)$ to the second, leaving the Attacker indifferent between attacking either of these two sites.

4.4 Let

$$p(S_i = 1 \mid d_i, a_i = 1) = 1 - d_i m_i$$

be the probability function for attack success at site i, where $0 \le m_i \le 1$ represents the effectiveness of the allocated defenses d_i in reducing this probability. Specifically, for each additional percentage point of allocated resources, there is a reduction of m_i in the probability of a successful attack on site i.

The Defender's expected loss is now

$$\tilde{\Psi}_D(d,a) = \mathbb{E}[v_1 S_1 + v_2 S_2] = v_1 (1 - d_1 m_1) a_1 + v_2 (1 - d_2 m_2) a_2.$$

The anticipated response of the attacker $a^*(d)$ is to prefer site i over j iff

$$v_i (1 - d_i m_i) > v_j (1 - d_j m_j).$$

The Defender minimizes her expected loss by allocating her resources as

$$d_1^* = \frac{v_1 + (m_2 - 1) v_2}{m_1 v_1 + m_2 v_2}, \quad d_2^* = \frac{(m_1 - 1) v_1 + v_2}{m_1 v_1 + m_2 v_2}.$$

In particular, if $m_1 = m_2 = 1$, one obtains the same solution as in Exercise 4.2.

If both sites have the same value, $v_1 = v_2$, the defender's optimal allocation is $d_1^* = m_2/(m_1 + m_2)$ and $d_2^* = m_1/(m_1 + m_2)$. For example, suppose $m_1 = 1$ and $m_2 = 0.5$, representing a 100% and 50% effectiveness of the allocated resources in protecting the first and second sites, respectively. The optimal allocation puts $d_1^* = 1/3$ of the resources on the site that will use them most effectively, and it puts $d_2^* = 2/3$ on the site where these resources are only half as effective. If the Defender were to transfer resources from the second to the first site, where they can be used more efficiently, the Attacker would respond by attacking the second site that is left with less protection, increasing her expected loss and, therefore, discouraging the Defender from doing so. Similar reasoning applies if the Defender were to move resources the other way around, from the first to the second site. This results from the minimax allocation d^* being an equilibrium, where the attacker is indifferent between both sites.

References

Alderson, D. L., Brown, G. G., Carlyle, W. M., and Wood, R. K. (2011). Solving Defender-Attacker-Defender Models for Infrastructure Defense. In *Operations Research, Computing, and Homeland Defense*, R. K. Wood and R. F. Dell, editors, pp. 28–49. INFORMS, Hanover, MD.

Aliprantis, C. D., and Chakrabarti, S. (2010). *Games and Decision Making*, 2nd ed. Oxford University Press, Oxford, U.K.

Alpern, S., and Gal, S. (2003). *The Theory of Search Games and Rendezvous*. Springer, New York, NY.

Amutio, M., Candau, J., and Mañas, J. A. (2014). *MAGERIT—version 3.0: Methodology for Information Systems Risk Analysis and Management*. Ministry of Finance and Public Administration, Madrid, Spain.

Arce, D., and Sandler, T. (2005). Counterterrorism: A Game-Theoretic Analysis. *Journal of Conflict Resolution*, **49**, 183–200.

Au, T. (2014). *Topics in Computational Advertising*, Ph.D. thesis, Duke University.

Aubin, J. P. (1993). *Optima and Equilibria*. Springer-Verlag, New York, NY.

Bajari, P. (2001). Comparing Competition and Collusion in Procurement Auctions: A Numerical Approach. *Economic Theory*, **18**, 187–205.

Banks, D., Petralia, F., and Wang, S. (2011). Adversarial Risk Analysis: Borel Games. *Applied Stochastic Models in Business and Industry*, **27**, 72–86.

Bayrak, H., and Bailey, M. D. (2008). Shortest Path Network Interdiction with Asymmetric Information. *Networks*, **52**, 133–140.

Bedford, T., and Cooke, R. (2001). *Probabilistic Risk Analysis: Foundations and Methods*. Cambridge University Press, Cambridge, U.K.

Bellman, R. (1969). On Colonel Blotto and Analogous Games. *SIAM Review*, **11**, 66–68.

Bellman, R., and Blackwell, D. (1949). Some Two-Person Games Involving Bluffing. *Proceedings of the National Academy of Sciences*, **35**, 600–605.

Berger, J., and Berliner, M. (1986). Robust Bayes and Empirical Bayes Analysis with ε-Contaminated Priors. *Annals of Statistics*, **14**, 461–486.

Bier V., and Azaiez N. (2009). *Game Theoretic Risk Analysis of Security Threats*. Springer, New York, NY.

Borel, E. (1938). *Traité du Calcul des Probabilités et ses Applications*, Vol. IV, Fascicule 2, *Applications aux Jeux des Hazard*. Gautier-Villars, Paris, France.

Brown, G., Carlyle, W., Salmeron, J., and Wood, K. (2006). Defending Critical Infrastructure. *Interfaces*, **36**, 530–544.

Brown G., Carlyle W., and Wood R. (2008). Optimizing Department of Homeland Security Defense Investments: Applying Defender-Attacker(-Defender) Optimization to Terror Risk Assessment and Mitigation. Appendix E of *Department of Homeland Security Bioterrorism Risk Assessment: A Call for Change*, National Academies Press, Washington, DC.

Brown, G. W. (1949). Some Notes on Computation of Games Solutions. Report P-78, The Rand Corporation, Santa Monica, CA.

Camerer, C. (2003). *Behavioral Game Theory*. Princeton University Press, Princeton, NJ.

Cardoso, J., and Diniz, P. (2009). Game Theory Models of Intelligent Actors in Reliability Analysis: An Overview of the State of the Art. In *Game Theoretic Risk Analysis of Security Threats*, pp. 1–19. Springer, New York, NY.

Carney, S. (2009). Cutthroat Capitalism: An Economic Analysis of the Somali Pirate Business Model. *Wired* (July 17), http://www.wired.com/politics/⊕ security/magazine/17-07/ff_somali_pirates. Accessed Sept. 23, 2014.

Case, J. (2008). *Competition*. Hill and Wang, NY.

Cho, I.-K., and Kreps, D. (1987). Signalling Games and Stable Equilibria. *Quarterly Journal of Economics*, **102**, 179–221.

Chu, H. (2010). Paris Metro's Cheaters Say Solidarity Is the Ticket. *Los Angeles Times*, June 22, articles.latimes.com. Accessed Dec. 12, 2014.

Clemen, R. T., and Reilly, T. (2004). *Making Hard Decisions with the Decision Tools Suite Update Edition*. Duxbury/Thomson Learning, Belmont, CA.

Clyde, M., and George, E. (2004). Model Uncertainty. *Statistical Science*, **19**, 81–94.

Couce Vieira, A., Houmb, S., and Ríos Insua, D. (2014). A Graphical Adversarial Risk Analysis Model for Oil and Gas Drilling Cybersecurity. In *Proceedings of the First International Workshop on Graphical Models for Security (GraMSec 2014)*, B. Kordy, S. Mauw, and W. Pieters, editors, pp. 78–93.

Cox, D. R. (1972). Regression Models and Life Tables. *Journal of the Royal Statistical Society, Series B*, **34**, 187–220.

Cox, Jr., L. A. (2008a). What's Wrong with Risk Matrices. *Risk Analysis*, **28**, 497–512.

Cox, Jr., L. A. (2008b). Some Limitations of "Risk = Threat × Vulnerability × Consequence" for Risk Analysis of Terrorist Attacks. *Risk Analysis*, **28**, 1749–1761.

Cox, Jr., L. A. (2013). *Improving Risk Analysis*. Springer Verlag, New York, NY.

Das, S., Yang, H., and Banks, D. (2013). Synthetic Priors that Merge Opinion from Multiple Experts. *Statistics, Politics and Policy*, **4**, DOI: 10.1515/215 ⊕ 1-7509.1060.

De Finetti, B. (1974). Foresight: Its Logical Laws, Its Subjective Sources (translation of the 1937 article in French). In *Studies in Subjective Probability*, H. E. Kyburg and H. E. Smokler, editors. Wiley, New York, NY.

Dimitrov, N., Michalopoulos, D., Morton, D., Nehme, M., Pan, F., Popova, E., Schneider, E., and Thoreson, G. (2011). Network Deployment of Radiation Detectors with Physics-Based Detection Probability Calculations. *Annals of Operations Research*, **187**, 207–228.

Dixit, A. (1980). The Role of Investment in Entry-Deterrence. *The Economic Journal*, **90**, 95–106.

Dyer, J. S., and Sarin, R. K. (1979). Measurable Multiattribute Value Functions. *Operations Research*, **27**, 810–822.

Dyer, J. and Sarin, R. (1982). Relative Risk Aversion. *Management Science*, **28**, 875–886.

English, R. (2009). *Terrorism*. Oxford University Press, Oxford, U.K.

Esteban, P. G., and Ríos Insua, D. (2014a). Supporting an Autonomous Social Agent within a Competitive Environment. *Cybernetics and Systems*, **43**, 241–253.

Esteban, P., and Ríos Insua, D. (2014b). From Cooperation to Competition. Technical Report, Universidad Rey Juan Carlos, Madrid, Spain.

Ezell, B., Bennett, S., von Winterfeldt, D., Sokolowski, J., and Collins, A. (2010). Probabilistic Risk Analysis and Terrorism Risk. *Risk Analysis*, **30**, 575–589.

Farquhar, P. (1984). State of the Art: Utility Assessment Methods. *Management Science*, **30**, 1283–1300.

Feller, W. (1968). An Introduction to Probability Theory and Its Applications, vol 1, 3rd ed. Wiley, New York, NY.

Ferguson, C., and Ferguson, T. S. (2003). On the Borel and von Neumann Poker Models. *Game Theory and Applications*, **9**, 17–32.

Ferguson, C., Ferguson, T. S., and Gawargy, C. (2007). Uniform(0,1) Two-Person Poker Models. *Game Theory and Applications*, **12**, 17–37.

Ferguson, T. S. (1973). A Bayesian Analysis of Some Nonparametric Problems. *The Annals of Statistics*, **1**, 209–230.

Fibich, G., and Gavious, N. (2011). Numerical Solutions of Asymmetric First-Price Auctions. *Games and Economic Behavior*, **73**, 479–495.

French, S., and Ríos Insua, D. (2000). *Statistical Decision Theory*. Arnold, London, U.K.

Garthwaite, P., Kadane, J. B., and O'Hagan, A. (2005). Statistical Methods for Eliciting Probability Distributions. *Journal of the American Statistical Association*, **100**, 680–701.

Gayle, W.-R., and Richard, J.-F. (2008). Numerical Solutions of Asymmetric, First-Price, Independent Private Values Auctions. *Computational Economics*, **32**, 245–278.

Ghosh, J. K., and Ramamoorthi, R. V. (2008). *Bayesian Nonparametrics*. Springer, New York, NY.

Gibbons, R. (1992). *Game Theory for Applied Economists*. Princeton University Press, Princeton, NJ.

Gintis, H. (2009). *The Bounds of Reason: Game Theory and the Unification of the Behavioral Sciences*. Princeton University Press, Princeton, N.J.

Good, I. J. (1953). The Population Frequencies of Species and the Estimation of Population Parameters. *Biometrika*, **40**, 237–264.

Haberfeld, M. and von Hassell, A. (2009). *A New Understanding of Terrorism: Case Studies, Trajectories and Lessons Learned*. Humanities, Social Sciences and Law series, Springer, New York, NY.

Haimes, Y. (2004). *Risk Modeling, Assessment and Management*. Wiley, Hoboken, NJ.

Harrington, J. (2014). *Games, Strategies and Decision Making*, 2nd ed. Worth Publishers, New York, NY.

Harsanyi, J. (1967a). Games with Incomplete Information Played by Bayesian Players, Part I. The Basic Model. *Management Science*, **14**, 159–182.

Harsanyi, J. (1967b). Games with Incomplete Information Played by Bayesian Players, Part II. Bayesian Equilibrium Points. *Management Science*, **14**, 320–334.

Harsanyi, J. (1968). Games with Incomplete Information Played by Bayesian Players, Part III. The Basic Probability Distribution of the Game. *Management Science*, **14**, 486–502.

Harsanyi, J. (1982). Subjective Probability and the Theory of Games: Comments on Kadane and Larkey's Paper. *Management Science*, **28**, 120–124.

Heal, G., and Kunreuther, H. (2007). Modeling Interdependent Risks. *Risk Analysis*, **27**, 621–634.

Henriques-Gomes, L. (2013). Twitter, Facebook Ticket Inspector Warnings for Fare Evaders. *Leader*, May 17, http://www.heraldsun.com.au/leader/. Accessed December 12, 2014.

Hjort, N. L., Holmes, C., Müller, P., and Walker, S. G. (2010). *Bayesian Nonparametrics*. Cambridge University Press, Cambridge, U.K.

Ho, T., Camerer, C., and Weigelt, K. (1998). Iterated Dominance and Iterated Best Response in Experimental "p-Beauty Contests." *The American Economic Review*, **88**, 947–969.

Hoeting, J., Madigan, D., Raftery, A., Volinsky, C. (1999). Bayesian Model Averaging: A Tutorial. *Statistical Science*, **14**, 382–417.

Holt, C., and Laury, S. (2002). Risk Aversion and Incentive Effects. *American Economic Review*, **92**, 1644–1655.

Hubbard, T. P., Kirkegaard, R., and Paarsch, H. J. (2012). Using Economic Theory to Guide Numerical Analysis: Solving for Equilibria in Models of Asymmetric First-Price Auctions. *Computational Economics*, **42**, 241–266.

International Organization for Standardization (2009). *ISO 31000:2009 Risk Management—Principles and Guidelines*. www.iso.org/iso/catalo ⊕ gue_detail.htm?csnumber=43170. Accessed on November 26, 2014.

Jain, M., Tsai, J., Pita, J., Kiekintveld, C., Rathi, S., and Tambe, M. (2010). Software Assistants for Randomized Patrol Planning for the LAX Airport Police and the Federal Air Marshal Service. *Interfaces*, **40**, 267–290.

Jittorntrum, K. (1978). An Implicit Function Theorem. *Journal of Optimization Theory and Application*, **25**, 575–577.

Kadane, J. B. (2009). Bayesian Thought in Early Modern Detective Stories: Monsieur Lecoq, C. Auguste Dupin and Sherlock Holmes. *Statistical Science*, **24**, 238–243.

Kadane, J. B., and Larkey, P. D. (1982). Subjective Probability and the Theory of Games. *Management Science*, **28**, 113–120. Reply: 124.

Kaelbling, L., and Littman, M. (1996). Reinforcement Learning: A Survey. Journal of Artificial Intelligence Research, **4**, 237–285.

Kahneman, D. (2003). Maps of Bounded Rationality: Psychology for Behavioral Economists. *American Economic Review*, **93**, 1449–1475.

Kahneman, D., Knetsch, J., and Thaler, R. (1990). Experimental Test of the Endowment Effect and the Coase Theorem. *Journal of Political Economy*, **98**, 1325–1348.

Kahneman, D., and Tversky, A. (1979). Prospect Theory: An Analysis of Decision under Risk. *Econometrica*, **47**, 263–291.

Kaplan, T., and Zamir, S. (2012). Asymmetric First-Price Auctions with Uniform Distributions: Analytic Solutions to the General Case. *Economic Theory*, **50**, 269–302.

Kardes, E. (2007). *Robust Stochastic Games and Applications to Counterterrorism Startegies*. Technical Report, CREATE, University of Southern California.

Karlin, S. (1959). *Mathematical Methods and Theory in Games, Programming and Economics*, vol. 1, reprinted in 1992. Dover Publications, New York, NY.

Karlin, S., and Restrepo, R. (1957). Multistage Poker Models. In *Contributions to Theoretical Games III*, M. Dresher, A. W. Tucker and P. Wolfe, editors, pp. 337–363. Princeton University Press, Princeton, NJ.

Keefer, D. (1991). Resource Allocation Models with Risk Aversion and Probabilistic Dependence: Offshore Oil and Gas Bidding. *Management Science*, **37**, 377-395.

Keeney, G., and von Winterfeldt, D. (2010). Identifying and Structuring the Objectives of Terrorists. *Risk Analysis*, **30**, 1803–1816.

Keeney, R. (2007). Modeling Values for Anti-Terrorism Analysis. *Risk Analysis*, **27**, 585–596.

Keeney, R., and Raiffa, H. (1993). *Decision Making with Multiple Objectives*. Cambridge University Press, Cambridge, U.K.

Keeney, R., and von Winterfeldt, D. (2011). A Value Model for Evaluating Homeland Security Decisions. *Risk Analysis*, **31**, 1470–1487.

Klemperer, P. (2004). *Auctions: Theory and Practice*. Princeton University Press, Princeton, NJ.

Koller, D., and Milch, B. (2003). Multi-Agent Influence Diagrams for Representing and Solving Games. *Games and Economic Behavior*, **45**, 181–221.

Kraan, B., and Bedford, T. (2005). Probabilistic Inversion of Expert Judgments in the Quantification of Model Uncertainty. *Management Science*, **51**, 995–1006.

Krishna, V. (2010). *Auction Theory*, 2nd ed. Elsevier, Oxford, U.K.

Kunreuther, H., and Heal, G. (2003). Interdependent Security. *Journal of Risk and Uncertainty*, **26**, 231–249.

Lebrun, B. (1996). Existence of an Equilibrium in First Price Auctions. *Economic Theory*, **7**, 421–443.

Lebrun, B. (1999). First Price Auctions in the Asymmetric *n* Bidder Case. *International Economic Review*, **40**, 125–142.

Lebrun, B. (2006). Uniqueness of the Equilibrium in First-Price Auctions. *Games and Economic Behavior*, **55**, 131–151.

Lee, R. and Wolpert, D. (2012). Game Theoretic Modeling of Pilot Behavior during Mid-Air Encounters. In *Decision Making with Imperfect Decision Makers*, T. V. Guy, M. Kárný, and D. Wolpert, editors, pp. 75–112. Springer Verlag, New York, NY.

Levine, B., Lu, A. and Reddy, A.V. (2013). Measurement of Subway Service Performance at New York City Transit. *Transportation Research Record: Journal of the Transportation Research Board*, **2353**, 57–68.

Li, H., and Riley, J. G. (2007). Auction Choice. *International Journal of Industrial Organization*, **25**, 1269–1298.

Luce, R. D., and Raiffa, H. (1957). *Games and Decisions: Introduction and Critical Survey*. Wiley, New York, NY.

Marshall, R. C., Meurer, M. J., Richard, J.-F., and Stromquist, W. (1994). Numerical Analysis of Asymmetric First-Price Auctions. *Games and Economic Behavior*, **7**, 193–220.

Martinez, J. E., and Mendez, I. (2009). Que Podemos Saber Sobre el Valor Estadistico de la Vida en Espana Utilizando Datos Laborales. *Hacienda Publica Espanola*, **190**, 73–93.

Maskin, E. S., and Riley, J. G. (2000a). Asymmetric Auctions. *The Review of Economic Studies*, **67**, 413–438.

Maskin, E. S., and Riley, J. G. (2000b). Equilibrium in Sealed High Bid Auctions. *The Review of Economic Studies*, **67**, 439–454.

McLay, L., Rothschild, C., and Guikema, S. (2012). Robust Adversarial Risk Analysis: A Level-k Approach. *Decision Analysis*, **9**, 41–54.

Menache, I., and Ozdaglar, A. (2011). *Network Games: Theory, Models and Dynamics*, Morgan Claypool, San Rafael, CA.

Merrick, J., and Parnell G. (2011). A Comparative Analysis of PRA and Intelligent Adversary Methods for Counterterrorism Management. *Risk Analysis*, **31**, 1488–1510.

Moreno, E., and Girón, J. (1998). Estimating with Incomplete Count Data: A Bayesian Approach. *Journal of Statistical Planning and Inference*, **66**, 147–159.

Müller, P., and Rodriguez, A. (2013). *Nonparametric Bayesian Inference*. NSF-CBMS Conference Series in Probability and Statistics, vol. 9. Institute of Mathematical Statistics, Beachwood, OH.

Myerson, R. (1991). *Game Theory: Analysis of Conflict*. Harvard University Press, Cambridge MA.

Nagel, R. (1995). Unraveling in Guessing Games: An Experimental Study. *American Economic Review*, **85**, 1313–1326.

Nash, J. (1951). Non-cooperative Games. *The Annals of Mathematics*, **54**, 286–295.

National Technical Authority for Information Assurance (2009). *HMG IA Standard No. 1: Technical Risk Assessment*. CESG, Cheltenham, U.K.

Nau, R. (1992). Joint Coherence in Games of Incomplete Information. *Management Science*, **38**, 374–387.

Newman, D. (1959). A Model for "Real" Poker. *Operations Research*, **7**, 557–560.

O'Hagan, A., Buck, C., Daneshkhah, A., Eiser, J. R., Garthwaite, P. H., Jenkinson, D. J., Oakley, J. E., and Rakow, T. (2006). *Uncertain Judgements: Eliciting Experts' Probabilities*. Wiley, Hoboken, NJ.

Osborne, M. (2004). *An Introduction to Game Theory*. Oxford University Press USA, New York, NY.

Parnell, G., Banks, D., Borio, L., Brown, G., Cox, Jr., L. A., Gannon, J., Harvill, E., Kunreuther, H., Morse, S., Pappaioanou, M., Pollack, S., Singpurwalla, N., and Wilson, A. (2008). *Report on Methodological Improvements to the Department of Homeland Security's Biological Agent Risk Analysis*. National Academies Press, Washington, D.C.

Parnell, G., Smith, C., and Moxley, F. (2010). Intelligent Adversary Risk Analysis: A Bioterrorism Risk Management Model. *Risk Analysis*, **30**, 32–48.

Paté-Cornell, E., and Guikema, S. (2002). Probabilistic Modeling of Terrorist Threats: A Systems Analysis Approach to Setting Priorities among Countermeasures. *Military Operations Research*, **7**, 5–23.

Pearl, J. (2005). Influence Diagrams: Historical and Personal Perspectives. *Decision Analysis*, **2**, 232–234.

Ploch, L., Blanchard, C. M., O'Rourke, R., Mason, R. C., and King, R. O. (2009). Piracy Off the Horn of Africa. U.S. Congressional Research Service, `http://fpc.state.gov/documents/organization/130809.p` ⊕ `df`. Accessed September 24, 2014.

Pratt, J. W. (1964). Risk Aversion in the Small and in the Large. *Econometrica*, **32**, 122–136.

Raiffa, H. (1982). *The Art and Science of Negotiation*. Harvard University Press, Cambridge, MA.

Raiffa, H., Richardson, J., and Metcalfe, D. (2002). *Negotiation Analysis: The Science and Art of Collaborative Decision*. Harvard University Press, Cambridge, MA.

Rapoport, A. (1970). *Two Person Game Theory: The Essential Ideas*. University of Michigan Press, Ann Arbor, MI.

Rázuri, J., Esteban, P. G., and Ríos Insua, D. (2013). An Adversarial Risk Analysis Model for an Autonomous Imperfect Decision Agent. In *Decision Making and Imperfection*, pp 163–187. Springer, Berlin, Germany.

Reddy, A. V., Kuhls, J., and Lu, A. (2011) Measuring and Controlling Subway Fare Evasion. *Transportation Research Record: Journal of the Transportation Research Board*, **2216**, 85–99.

Reilly, B., Rickman, N., and Witt, R. (2012). Robbing Banks: Crime Does Pay—But Not Very Much. *Significance*, **9**, 17–21.

Rios, J., and Ríos Insua, D. (2012). Adversarial Risk Analysis for Counterterrorism Modeling. *Risk Analysis*, **32**, 894–915.

Ríos Insua, D. (1990). *Sensitivity Analysis in Multiobjective Decision Making*. Springer, Berlin, Germany.

Ríos Insua, D., Cano, J., Pellot, M. and Ortega, R. (2015). Current Trends in Bayesian Methodology with Applications. In *From Risk Analysis to Adversarial Risk Analysis* (to appear). CRC Press, Boca Raton, FL.

Rios Insua, D., Kersten, G., Rios, J., and Grima, C. (2008). Towards Decision Support for Participatory Democracy. *Information Systems and e-Business Management*, **6**, 161–191.

Ríos Insua, D., Rios, J., and Banks, D. (2009). Adversarial Risk Analysis. *Journal of the American Statistical Association*, **104**, 841–854.

Ríos Insua, D., and Ruggeri, F. (2000). *Robust Bayesian Analysis*. Springer, Berlin, Germany.

Roberson, B. (2006). The Colonel Blotto Game. *Economic Theory*, **29**, 1–24.

Rothkopf, M. H. (2007). Decision Analysis: The Right Tool for Auctions. *Decision Analysis*, **4**, 167–172.

Rothschild, C., McLay, L., and Guikema, S. (2012). Adversarial Risk Analysis with Incomplete Information: A Level-k Approach. *Risk Analysis*, **32**, 1219–1231.

Rubinstein, A. (1998). *Modeling Bounded Rationality*. MIT Press, Cambridge, MA.

Sakaguchi, M. (1984). A Note on the Disadvantage for the Sente in Poker. *Mathematica Japonica*, **29**, 483–489.

Sakaguchi, M., and Sakai, S. (1981). Partial Information in a Simplified Two-Person Poker. *Mathematica Japonica*, **26**, 695–705.

Samuelson, P. A. (1977). St. Petersburg Paradoxes: Defanged, Dissected, and Historically Described. *Journal of Economic Literature*, **15**, 24–55.

Sasaki, Y. (2014). Optimal Choices of Fare Collection Systems for Public Transportations: Barrier Versus Barrier-Free. *Transportation Research Part B: Methodological*, **60**, 107–114.

Savage, L. J. (1954). *The Foundations of Statistics*. Wiley, New York, NY.

Seidenfeld, T., Kadane, J. B., and Schervish, M. (1989). One the Shared Preferences of Two Bayesian Decision Makers. *The Journal of Philosophy*, **86**, 225–244.

Selman, B., Kautz, H. A., and Cohen, B. (1994). Noise Strategies for Improving Local Search. In Proceedings of the Twelfth National Conference on Artificial Intelligence, pp. 337–343. AAAI Press, Palo Alto, CA.

Sevillano, J. C., Ríos Insua, D., and Rios, J. (2012). Adversarial Risk Analysis: The Somali Pirates Case. *Decision Analysis*, **9**, 81–85.

Shachter, R. (1986). Evaluating Influence Diagrams. *Operations Research*, **34**, 871–882.

Shan, X., and Zhuang, J. (2014). Modeling Credible Retaliation Threats in Deterring the Smuggling of Nuclear Weapons Using Partial Inspections: A Three-Stage Game. *Decision Analysis*, **11**, 43–62.

Simon, H. (1955). A Behavioral Model of Rational Choice. *Quarterly Journal of Economics*, **69**, 99–118.

Singpurwalla, N. (2010). *Reliability and Risk*. Wiley, Hoboken, NJ.

Smith, M., and Clarke, R. (2000). Crime and Public Transport. *Crime and Justice*, **27**, 169–233.

Spence, A. M. (1974). *Market Signalling: Information Transfer in Hiring and Related Processes*. Harvard University Press, Cambridge, MA.

Stahl, D., and Wilson, P. (1994). Experimental Evidence on Players' Models of Other Players. *Journal of Economic Behavior and Organization*, **25**, 309–327.

Stahl, D., and Wilson, P. (1995). On Players' Models of Other Players: Theory and Experimental Evidence. *Games and Economic Behavior*, **10**, 218–254.

Stoneburner, G., Goguen, A., and Feringa, A. (2002) *Risk Management Guide for Information Technology Systems*. National Institute of Standards and Technology Special Publication 800-30. NIST, Gaithersburg, MD.

Thomson, W. (1994). Cooperative Models of Bargaining. In *Handbook of Game Theory with Economic Applications*, R. Aumann and S. Hart, editors, pp. 1237–1284. North-Holland, Amsterdam, NL.

Troelsen, L., and Barr, L. (2012). Combating Pickpocketing in Public Transportation. *Public Transport International*, **61**, 32–33.

Vickrey, W. (1961). Counterspeculation, Auctions, and Competitive Sealed Tenders. *The Journal of Finance*, **16**, 8–37.

Viscusy, K., and Aldy, J. (2003). The Value of a Statistical Life: A Critical Review of Market Estimates throughout the World. *Journal of Risk and Uncertainty*, **27**, 5–76

von Neumann J., and Morgenstern, O. (1944). *Theory of Games and Economic Behavior*. Princeton University Press, Princeton, NJ.

von Winterfeldt D., and O'Sullivan T. M. (2006). Should We Protect Commercial Airplanes against Surface-to-Air Missile Attacks by Terrorists? *Decision Analysis*, **3**, 63–75.

Wakker, P. (2010) *Prospect Theory: For Risk and Ambiguity*. Cambridge University Press, Cambridge, U.K.

Wang, C., and Bier, V. (2012). Optimal Defensive Allocations in the Face of Uncertain Terrorist Preferences, with an Emphasis on Transportation. *Homeland Security Affairs*, www.hsaj.org/?special:fullarticle=0.4.4. Accessed Dec. 27, 2014.

Wang, C., and Bier, V. (2013). Expert Elicitation of Adversary Preferences Using Ordinal Judgments. *Operations Research*, **61**, 372–385.

Wang, S., and Banks, D. (2011). Network Routing for Insurgency: An Adversarial Risk Analysis Framework. *Naval Research Logistics*, **58**, 595–607.

Washburn, A., and Wood, R. K. (1995). Two-Person Zero-Sum Games for Network Interdiction. *Operations Research*, **43**, 243–251.

Xu, J., and Zhuang, J. (2015). Modeling Costly Learning and Counter-Learning in an Attacker-Defender Game with Private Defender Information. *Annals of Operations Research*, forthcoming.

Zhuang, J., Bier, V., and Alagoz, O. (2005). Modeling Secrecy and Deception in a Multiple-Period Attacker-Defender Signaling Game. *European Journal of Operations Research*, **203**, 409–418.

Index

Printed in the United States
by Baker & Taylor Publisher Services